SPRINGER
LAB MANUAL

F. Francuskiewicz

# Polymer Fractionation

With 32 Figures and 42 Tables



Springer-Verlag
Berlin Heidelberg New York London Paris
Tokyo Hong Kong Barcelona Budapest

Dr. rer. nat. FRIEDER FRANCUSKIEWICZ
FARU, Forschungsstelle Dr. Kubsch
Laboratorium für Analytik, Radiometrie
und Umwelttechnologie GmbH
Hohe Straße 48
D-01187 Dresden, FRG

Editors for polymer science titles: Prof. Dr. rer. nat. habil. Glöckner, Papstdorfer Str. 25, 01277 Dresden and Dr. Howard Barth, E.I. du Pont de Nemours, Central Research and Development, P.O. Box 80228, Wilmington, DE 19880-0228, USA. This title has been edited by Prof. Glöckner.

ISBN 3-540-57539-1 Springer-Verlag Berlin Heidelberg New York
ISBN 0-387-57539-1 Springer-Verlag New York Berlin Heidelberg

Library of Congress Cataloging-in-Publication Data. Francuskiewicz, F. (Frieder), 1944 – Polymer fractionation/ F. Francuskiewicz. p. cm. – (Springer laboratory) Includes bibliographical references and index. ISBN 3-540-57539-1 (Berlin: acid-free). – ISBN 0-387-57539-1 (New York: acid-free) 1. Polymers – Separation. I. Title. II. Series. QD381.9.S44F73 1994 547.7'046–dc20



Typesetting: Macmillan India Ltd., Bangalore
2/3130-5 4 3 2 1 - Printed on acid-free paper

# Preface

Fractionation of polymers via solubility has been a well known method in polymer characterization for a long time. The original object of analytical fractionations, the determination of the molecular weight distribution, is nowadays achieved more efficiently and conveniently by chromatographic methods. However, fractionation procedures, which were developed in great diversity, remain up-to-date and essential for obtaining preparative fractions with narrow distributions. Such fractions are wanted increasingly for the investigation of true structure–property relationships which are mostly influenced by distributions of molecular weight or other parameters such as branching or chemical composition.

Literature on the field of polymer fractionation is extensive and several reviews exist. However, there is a lack of systematically methodical instructions for carrying-out of diverse fractionation procedures. This volume represents an attempt to reduce this deficiency and is focussed on practical aspects of fractionation procedures.

This laboratory manual is intended for polymer chemists, physicists, and technicians, for students of polymer science, and skilled laboratory assistants, all of whom are not dealing directly with fractionation but are in need of fractions to carry out further investigations.

The book is arranged into introductory and fundamental sections (1–4), followed by descriptions of the various fractionation procedures according to the conventional classification: precipitation fractionation (Sect. 5), extraction fractionation (Sect. 6), gradient-elution fractionation (Sect. 7), Baker-Williams fractionation (Sect. 8), temperature rising elution fractionation (Sect. 10), partition fractionation (Sect. 11), and cross fractionation (Sect. 12). Section 9 summarizes trouble shooting and problems with large-scale procedures which commonly occur in the fractionation techniques described in Sects. 6 to 8. Of course, due to overlapping of several procedures, a certain arbitrariness could not be completely avoided.

One procedure based on solubility, viz. supercritical fluid chromatography, is not dealt with because it requires very special and extensive equipment, experience, and skill. This method should be reserved for specialists.

Sections 5 to 11 (excluding Sect. 9) are each divided into a very short exposition of theoretical aspects (Principles and Limitations of Application), restricted to the knowledge necessary for understanding the practical procedures. This is followed by Equipment and Materials and by the discussion of

Preparatory Investigations and Fractionation Steps. Each procedure is illustrated by at least one practical example of carrying-out the fractionation. These examples are given as instructions for fractionations using apparatus which is as simple as possible. It can be easily varied or modified.

Besides fractionation according to molecular weight, separation with respect to chemical composition of copolymers (Sects. 4.4, 11.1, 12) and short-chain branching (Sects. 10, 12) are also discussed. Modern procedures of fractionation such as continuous polymer fractionation (Sect. 6.6) and fractionation with demixing solvents (Sect. 11.1) are also included.

According to the character and range of the present volume, complete reviews of the literature on the methods discussed are not given; references to reviews are listed in Appendix A 10. Not every fractionation problem could be taken into consideration, but small variations of the discussed procedures should make it possible to solve new fractionation problems. Appendices and Glossary offer supplementary or more detailed information, if required. To make the book convenient to use, some repetition, especially in described examples, and references to other sections are deliberately given in the text.

I wish to thank, in the first place, the editor of this book, Prof. Dr. G. Glöckner, who has given continual support and many constructive suggestions. I am also obliged to Prof. Dr. H. W. Kammer for many fruitful discussions. The wealth of information kindly supplied by Prof. Dr. B. A. Wolf, Dr. R. Kuhn, Dr. J. Podešva, and Dr. J. Stejskal is gratefully acknowledged. Last but not least, thanks belong to the Springer-Verlag, especially Dr. Marion Hertel and the copyeditor, Mr. E. Fulford, for the pleasant cooperation.

Finally, I hope that this laboratory manual can fulfill the reader's expectations. Critical comments will be gratefully received.

FRIEDER FRANCUSKIEWICZ

# Contents

## Symbols and Abbreviations

| | |
|---|---|
| $a$ | exponent in the Kuhn-Mark-Houwink-Sakurada equation (Glossary) |
| AA | acetic acid |
| Ac | acetone |
| AcN | acetonitrile |
| AN | acrylonitrile |
| | |
| b.p. | boiling point |
| B | benzene |
| BD | butadiene |
| BuCl | *n*-butyl chloride |
| | |
| $c$ | concentration |
| CA | cellulose acetate |
| CC | chemical composition (of copolymers) |
| CCD | distribution of chemical composition |
| CF | chloroform, trichloromethane |
| CH | cyclohexane |
| CHN | cyclohexanone |
| CPE | chlorinated poly(ethylene) |
| CPF | continuous polymer fractionation |
| | |
| $d$ | diameter |
| $D$ | density |
| DBP | dibutyl phthalate |
| DCE | dichloroethane |
| DCM | dichloromethane |
| DHN | decahydronaphthalene, decaline |
| DMF | *N*,*N*-dimethyl formamide |
| DMSO | dimethyl sulfoxide |
| DP | degree of polymerization (in text) |
| DPD | distribution of polymerization degrees |
| | |
| $e_n$ | efficiency value of a fractionation (Eqs. (4.2), (4.3)) |
| $\Delta E$ | internal heat of evaporation |

| | |
|---|---|
| EA | extracting agent |
| EMA | ethyl methacrylate |
| EtA | ethyl acrylate |
| EtE | diethyl ether |
| EtOH | ethanol |
| EVA | poly(ethylene-*co*-vinyl acetate) |
| | |
| $f_n$ | efficiency value of a fractionation (Eqs. (4.4), (4.5)) |
| FA | fumaric acid |
| FD | feed |
| | |
| $G$ | ratio of polymer masses in sol and gel phase in CPF ($= m'_P/m''_P$) |
| $\dot{G}$ | ratio of mass fluxes of polymer in sol and gel phase in CPF ($= \dot{m}'_P/\dot{m}''_P$) |
| $\Delta G_{mix}$ | free energy of mixing |
| GL | gel |
| | |
| $H$ | heterogeneity of chain length ($= \bar{M}_w/\bar{M}_n = \bar{P}_w/\bar{P}_n$) |
| $\Delta H_{mix}$ | enthalpy of mixing |
| HDPE | high-density poly(ethylene) |
| HIP | hydrogenated isoprene |
| Hp | *n*-heptane |
| Hx | *n*-hexane |
| | |
| I($M$) | integral (cumulative) mass distribution function of $M$ |
| I($P$) | integral (cumulative) mass distribution function of $P$ |
| iOct | *iso*-octane, 2,2,4-trimethylpentane |
| iPrOH | *iso*-propanol |
| | |
| $k$ | proportionality parameter in Eqs. (3.8)–(3.11) depending on $\chi$ ($k \propto \chi$) |
| $K_i$ | distribution coefficient of species i |
| $K_v$ | constant factor in the Kuhn-Mark-Houwink-Sakurada equation (Glossary) |
| | |
| $l$ | length (of a column) |
| LDPE | low-density poly(ethylene) |
| LLDPE | linear low-density poly(ethylene) |
| LND | logarithmic normal distribution |
| | |
| $m_P$ | mass of polymer (CPF) |
| $\dot{m}_P$ | mass flux of polymer (CPF) |
| $M$ | molecular weight (in equations) |

| | |
|---|---|
| $M^*$ | median value of $M$ in LND (Table A 7) |
| $M_i$ | molecular weight of individual constituents i in a polymer |
| $\bar{M}_n$ | number average of $M$ |
| $M_0$ | molecular weight of the repeat unit |
| $\bar{M}_v$ | viscosity average of $M$ |
| $\bar{M}_w$ | weight average of $M$ |
| MA | maleic acid |
| MCH | methyl cyclohexane |
| MEK | methyl ethyl ketone, 2-butanone |
| MEMA | 2-methoxyethyl methacrylate |
| MeOH | methanol |
| MMA | methyl methacrylate |
| MMA/EMA | poly(methyl methacrylate-*co*-ethyl methacrylate) |
| MW | molecular weight (in text) |
| MWD | molecular weight distribution |
| | |
| $n_i$ | number of moles of individual constituents i |
| $n_0$ | number of moles of monomers necessary to synthesize the mass of a polymer sample |
| N($M$) | (differential) frequency distribution function of $M$ |
| N($P$) | (differential) frequency distribution function of $P$ |
| NM | nitromethane |
| | |
| oDCB | *ortho*-dichlorobenzene |
| | |
| $P$ | degree of polymerization (in equations) |
| $P^*$ | median value of $P$ in LND (Table A 7) |
| $P_i$ | degree of polymerization of individual constituents i in a polymer |
| $\bar{P}_n$ | number average of $P$ |
| $\bar{P}_v$ | viscosity average of $P$ |
| $\bar{P}_w$ | weight average of $P$ |
| PAN | poly(acrylonitrile) |
| PBD | poly(butadiene) |
| PDMS | poly(dimethyl siloxane) |
| PE | poly(ethylene) |
| PEMA | poly(ethyl methacrylate) |
| PIB | poly(isobutylene) |
| PMMA | poly(methyl methacrylate) |
| Pn | *n*-pentane |
| POAlk | poly(oxyalkylene) |
| POE | poly(oxyethylene) |
| POP | poly(oxypropylene) |
| PP | poly(propylene) |

| | |
|---|---|
| PrOH | *n*-propanol |
| PS | poly(styrene) |
| PVAC | poly(vinyl acetate) |
| PVAL | poly(vinyl alcohol) |
| | |
| R | universal gas constant |
| | |
| S | styrene |
| $\Delta S_{mix}$ | entropy of mixing |
| S/AN | poly(styrene-*co*-acrylonitrile) |
| S/BD | poly(styrene-*co*-butadiene) |
| SCB | short-chain branching |
| SEC | size-exclusion chromatography |
| S/FA | poly(styrene-*co*-fumaric acid) |
| S/HIP | poly(styrene-*co*-hydrogenated isoprene) |
| SL | sol |
| S/MA | poly(styrene-*co*-maleic acid) |
| S/MEMA | poly(styrene-*co*-2-methoxyethyl methacrylate) |
| S/MMA | poly(styrene-*co*-methyl methacrylate) |
| | |
| $t$ | time |
| $T$ | temperature |
| T | toluene |
| TCB | 1,2,4-trichlorobenzene |
| TCE | trichlorocthane |
| TeCE | tetrachloroethane |
| TeCM | tetrachloromethane, carbon tetrachloride |
| THF | tetrahydrofurane |
| TREF | temperature rising elution fractionation |
| | |
| $U$ | non-uniformity according to Schulz ( $=H-1$) |
| | |
| $V$ | volume or molar volume |
| $\dot{V}$ | flow rate |
| $V_e$ | elution volume (SEC) |
| $V_m$ | volume of mixing vessel |
| VAC | vinyl acetate |
| VAL | vinyl alcohol |
| | |
| $w_i$ | weight fraction or mass of individual constituents i; copolymer composition (weight fraction of monomer unit i) |
| $w_0$ | mass of monomers necessary to synthesize the mass of a polymer sample |
| W | water |

| | |
|---|---|
| W($M$) | differential mass distribution function of $M$ |
| W($P$) | differential mass distribution function of $P$ |
| WP | working point |
| | |
| $x_i$ | copolymer composition (mole fraction of monomer unit i) |
| $X$ | composition of a mixture of molecules having different $P_i$ |
| | |
| αCN | α-chloronaphthalene |
| αMS | α-methylstyrene |
| $\delta$ | solubility parameter (Eq. (3.3)) |
| $\Delta$ | difference in a quantity |
| $[\eta]$ | intrinsic viscosity of a polymer solution |
| $\kappa$ | fractionation parameter (Eqs. (4.7), (4.8)) |
| $\Delta\mu$ | difference in chemical potential |
| $\sigma$ | fractionation parameter (Eq. (4.7)) |
| $\sigma$ | standard deviation of LND (Table A 7) |
| $\varphi$ | volume fraction |
| $\varphi^*$ | volume fraction of nonsolvent on cloud point or precipitation point (turbidimetric titration) |
| $\varphi_{m,t}$ | volume fraction of solvent or nonsolvent in the mixing vessel at time $t$ (Eq. (7.1)) |
| $\varphi_{st}$ | ditto in the storage vessel (Eq. (7.1)) |
| $\chi$ | Huggins constant |

*Subscripts:*

| | |
|---|---|
| A, B | monomer units in a copolymer, or solvent and nonsolvent |
| EA | extracting agent |
| FD | feed |
| NS | nonsolvent |
| P | polymer |
| S | solvent |
| WP | working point |
| ′ | sol phase (single primed) |
| ″ | gel phase (double primed) |

# 1 Introduction

## 1.1 The Structure of Polymers

All synthetic and some natural polymers (e.g. natural rubber or cellulose) are heterogeneous products in different respects. The variety of molecules in a certain polymer comprises differences in molecular structure as well as in size of molecules. These inhomogeneities are a result of reaction statistics in the synthesis of polymers.

Important features of polymer structure are *constitution*, *configuration* and *conformation* of molecules on the one hand and *molecular weight* on the other hand. A schematical survey of structural parameters of macromolecules is given in Table 1.1 (The parentheses contain some additional remarks, hints or examples). For more details, see textbooks on polymer science (for problems of molecular weight see also Sect. 2).

Structural parameters of macromolecules influence properties and applications of polymers. Therefore, it is important to obtain information about the structural characteristics of the macromolecules and their distribution in a polymer sample. The most important characteristics are *molecular weight*, *branching* and, for copolymers, *chemical composition*, *composition heterogeneity* and, perhaps, a connection between molecular weight (MW) and chemical composition (CC). Such information can be obtained from fractionation procedures based on different solubilities of the individual species in the polymer. The parameters listed in Table 1.1 influence the behavior of polymers in fractionation techniques. Let us mention a few examples of these influences.

In a copolymer sample, CC of the individual molecules influences the solubility. Long chain branching reduces the solubility compared to linear molecules with identical MWs. The solubility decreases when MW increases (which is the basis of the fractionation techniques discussed in this book). Polymers with the same CC may possess different molecular weight distributions (MWD) and averages, respectively (see Sect. 2). This leads to different solubilities.

Other parameters in Table 1.1 may also influence more or less the solubility via supermolecular or ordered structures caused by special kinds of constitution, configuration and conformation. For example, alternating and block copolymers often develop ordered structures (pre-crystalline structures, aggregates, clusters) in solution. Likewise, *tacticity* (syndiotactic and isotactic

**Table 1.1.** Structural parameters of macromolecules

| Constitution | Configuration | Conformation | Molecular size |
|---|---|---|---|
| (Arrangement of monomer units one another)<br>– Head-to-tail (1,2-) structure:<br>• undisturbed (ionic polymerization of vinyl polymers)<br>• disturbed with head-to-head (1,1-) structure (radical polymerization of vinyl polymers)<br>– Diene structures:<br>• 1,2-polydiene<br>• 1,4-polydiene (polybutadiene, dependent on polymerization conditions)<br>– Branched structures:<br>• no branching (linear molecules)<br>• long chain branching (intermolecular chain transfer)<br>• short chain branching (intramolecular chain transfer)<br>– Copolymers:<br>• chemical heterogeneity (chemical composition distribution)<br>• sequence distribution of monomer units (random, alternating, block copolymers) | (Steric arrangement of substituents or ligands)<br>– *cis-trans* isomerism: (macromolecules containing double bonds, similar problems are present in molecules with ring structures)<br>– Tacticity:<br>• atactic (random arrangement of substituents, typical for ordinary polymers)<br>• syndiotactic } (stereoregular<br>• isotactic }<br>structures synthesized by special techniques like insertion polymerization) | (Rotation isomerism – *trans* and *gauche* conformation)<br>– All-*trans* conformation:<br>→ planar zigzag chain (ordered structure)<br>– All-*gauche* conformation:<br>→ helix (ordered structure)<br>– *trans-gauche* (regular change):<br>→ (extended) helix (ordered structure)<br>– *trans-guache* (random change):<br>→ random coil (mainly in solution and in amorphous polymers) | (Molecular weight and degree of polymerization, respectively)<br>– Molecular weight distribution (distribution of degree of polymerization)<br>– Average molecular weight (average degree of polymerization)<br>(Details see Sect. 2) |

structures) and conformation (zigzag or helix chains) may act in the same way. Sometimes, it depends on the "history" of the polymer (conditions of synthesis, ageing, temperature effects) to which extend such ordered structures become operative in a solution.

The probability to find two identical macromolecules within a synthetic polymer sample is extremely small. That means, on the other hand, that each polymer sample differs from another one more or less even for the same CC.

## 1.2 The Need for Fractionations and Their Scope

The fractionation of polymers is a tool for measuring the distribution of structural parameters. Mostly, the determination of MWD or the dependence of other molecular parameters and properties upon MW are in the foreground. Therefore, this book is chiefly focussed on fractionation with respect to MW without ignoring other factors totally.

In a fractionation procedure, the original polymer will be subdivided into fractions, mostly with different MWs and MWDs narrow in comparison to the starting polymer. These fractions must be analyzed with special methods to determine MWs and other structural features.

About twenty-five years ago, fractionation procedures were mainly the method to measure MWD. Nowadays, analytical fractionations are obsolete because newer, more efficient and convenient methods, first and foremost size-exclusion chromatography (SEC), are available to this end [1–3]. Nevertheless, the necessity of carrying out fractionations still exists since fractions of the same polymer sample graded in MW and having MWDs as narrow as possible are necessary for various investigations. Many parameters or properties in polymer characterization as well as in application of polymers depend on MW. To determine such dependencies, one often needs fractions with masses of a few grams to perform some other measurements in order to determine the desired relation. Well known examples are the calibration of SEC or viscometric measurements of fractions having known MWs for the evaluation of the parameters of the Kuhn-Mark-Houwink-Sakurada equation (see Glossary). The determination of the influence of MW on melt viscosity or mechanical properties of the solid polymer also requires fractions. The problem can be solved by fractionation on a preparative scale.

Of course, preparative SEC is also an established method [2]. However, in one run and on lab columns about 6 cm in diameter, about one gram of polymer can be fractionated. That means that for large fractions the corresponding fractions of many SEC runs must be combined together and this requires an excellent reproducibility of the method and a large amount of solvent. Furthermore, the costs for preparative SEC procedures are much higher than for preparative fractionations using differences in the solubility of polymers [2].

A second (and secondary) aspect for the realization of fractionations is that the solution behavior of polymers in solvent/nonsolvent systems or at different temperatures can be checked using fractionation procedures. For instance, information about the influence of MW and CC on solubility can be derived from results of fractionations in different solvent/nonsolvent mixtures. For these questions, fractionations on an analytical scale are sufficient.

In principle, preparative fractionations are large-scale experiments which are more time-consuming and expensive than analytical ones. It is advisable to start any preparative fractionation only after detailed knowledge has been gained about the behavior of the fractionation system and some experience and skill in the fractionation technique chosen have been acquired by fractionation experiments on an analytical scale. For this reason, this laboratory manual puts special emphasis on analytical aspects of fractionation.

The basic principle of fractionation techniques is the subdivision of the polymer sample into fractions due to the different solubility of the species. In other words, one can perform a fractionation by changing the dissolution power. This can be achieved in different ways using both solvent/nonsolvent systems and temperature variation:

- *Decrease of dissolution power* can be reached by enrichment of the nonsolvent in a solvent/nonsolvent mixture or by changing (mostly decrease) of temperature. With these procedures a *precipitation fractionation* (Sect. 5) is carried out. In its course, the first or last fractions contain molecules with the highest or lowest MWs, respectively.
- *Increase of dissolution power* leads to the result that molecules with the lowest MWs represent the first fraction and vice versa. Increase of solubility can happen by a stepwise or a continuous increase of the amount of solvent in the fractionation system or by a rise in temperature. The stepwise procedure is the *extraction fractionation* (Sect. 6) whereas methods with continuously increasing dissolution power often are called *gradient* or *elution fractionation* (Sect. 7). With a solvent/nonsolvent gradient, one has the *solvent-gradient elution fractionation.* Combinations with temperature programs in a different manner are also possible. A special version is the *precipitation chromatography according to Baker and Williams* (Sect. 8). The *temperature rising elution fractionation* (Sect. 10) works only with temperature programs.
- Differences in dissolution power can be achieved by combination of demixing solvents. In this way a *fractionation by partition* between demixing liquids (Sect. 11) is possible.

If copolymers are investigated, fractionation can be additionally influenced by heterogeneities in CC, i.e., fractions are different in both MW and CC. Sometimes, one is able to diminish or to intensify this effect by use of different solvent/nonsolvent systems (see Sect. 4.4).

Finally, one should note that no fractionation resembles another one completely, even if samples of the same type of polymer, e.g., poly(styrene), are fractionated by the same procedure. This is due to the variety of structural characteristics in the polymer samples (as mentioned in Sect. 1.1) and to the possibility that these features may be coupled (for instance, increasing MW with increasing long-chain branching).

## References

1. Tung LH, Moore JC (1977) In: Tung LH (ed) Fractionation of synthetic polymers. Marcel Dekker, New York, p 545
2. Glöckner G (1987) Polymer characterization by liquid chromatography. Elsevier, Amsterdam
3. Dawkins JV (1989) In: Booth C, Price C (eds) Comprehensive polymer science, vol 1. Pergamon, Oxford, p 231

# 2 The Size of Polymer Molecules – Mean Values and Distributions of the Molecular Weight

The size of a macromolecule can be described by the molecular weight (MW) $M$ as well as by the degree of polymerization (DP) $P$. Both quantities are related according to

$$M = M_0 \cdot P \tag{2.1}$$

where $M_0$ represents the MW of the repeat unit. Mostly, one applies MW to characterize the size of macromolecules but sometimes use of DP is advantageous, e.g., in the comparison of molecular size before and after a polymer-analogous reaction.

MWD, which is inherent to synthetic polymers, originates from randomness of the primary (elementary) steps in polymerization, polyaddition, and polycondensation reactions related to a single macromolecule.

Averages of MW can be calculated with help of sum expressions under consideration of the frequency of each MW in the sample [1–3]. The amount of the species is expressed either by the number of moles $n_i$ of molecules with $M_i$ or as the mass $w_i$ of this species. This leads to the *number average* $\bar{M}_n$ and the weight average $\bar{M}_w$, respectively:

$$\bar{M}_n = \sum (n_i M_i) / \sum n_i = \sum w_i / \sum (w_i / M_i) = \bar{P}_n \cdot M_0 \tag{2.2}$$

$$\bar{M}_w = \sum (n_i M_i^2) / \sum (n_i M_i) = \sum (w_i M_i) / \sum w_i = \bar{P}_w \cdot M_0 \tag{2.3}$$

Equation (2.1) can be rewritten as $M_i = M_0 P_i$, thus $n_i M_i = w_i = n_i P_i \cdot M_0$. Number averages can be measured [2, 4] by colligative methods such as osmotic or vapor pressure measurements and by end-group determinations. Weight averages are measured mainly by light-scattering methods [2, 5–7]. Viscometry, which is frequently used in MW determination, yields the so called *viscosity average* $\bar{M}_v$ [8–10]:

$$\begin{aligned} \bar{M}_v &= \{\sum (n_i M_i^{a+1}) / \sum (n_i M_i)\}^{1/a} \\ &= \{\sum (w_i M_i^a) / \sum w_i\}^{1/a} = \bar{P}_v \cdot M_0 \end{aligned} \tag{2.4}$$

In Eq. (2.4), $a$ is the exponent in the Kuhn-Mark-Houwink-Sakurada equation (see Glossary). Equations (2.2–2.4) can be used for the calculation of average values from fractionation data. In this case, $n_i$ is the number of moles with average $M_i$ and weight fraction $w_i$ in fraction i. All averages just mentioned are available from SEC analysis [11–13] where they are calculated after slicing the peak of the sample under consideration.

Number and weight averaged MWs give together a first estimate of the width of a distribution. Common quantities are the *MW heterogeneity*

$$H = \bar{M}_w/\bar{M}_n \geqslant 1 \tag{2.5}$$

and the *non-uniformity* according to Schulz

$$U = H - 1 \geqslant 0 \tag{2.6}$$

Fractionation according to MW should yield fractions with values of $H$ and $U$ smaller than the related quantities of the starting polymer.

Owing to the large number of species $M_i$ in a real polymer, it is possible and convenient to transform the sum expressions into continuous distributions without a noticeable error [1–3, 14, 15]. MWD covers all individual species of macromolecules within a polymer sample and specifies the amount of molecules with a certain MW applied to all molecules in the sample. Polymer fractionations serve (among other things) to obtain information about these amounts of different macromolecules in a polymer sample, i.e., about MWD.

Such a distribution must be discontinuous because the difference in two neighbouring species is

$$\Delta M_i = M_i - M_{i-1} = M_0 \tag{2.7}$$

Nevertheless, discontinuous distributions can be substituted by continuous functions when $M_0$ is very much less than the range of all $M_i$.

The value of the *frequency distribution* $\mathrm{N}(M)$ for macromolecules with $M_i$ is

$$\mathrm{N}(M_i) = n_i/n_0 \ll 1 \tag{2.8}$$

where $n_0 = \sum(n_i P_i)$ is the number of moles of repeat units in the whole polymer containing the molecules with $M_i$. Consequently, $\mathrm{N}(M)$ is related to one mole of repeat units.

For the *mass distribution* $\mathrm{W}(M)$ one obtains by analogy:

$$\mathrm{W}(M_i) = w_i/w_0 \ll 1 \tag{2.9}$$

$w_0$ in Eq. (2.9) is the mass of monomer reacted to the same mass of polymer, i.e., $\mathrm{W}(M)$ is related to one gram of repeat units (which is equal to one gram of polymer).

By comparison of Eqs. (2.8) and (2.9) it follows:

$$\mathrm{W}(M_i) = n_i \cdot P_i \cdot M_0/(n_0 \cdot 1 \cdot M_0) = P_i \cdot \mathrm{N}(M_i) \tag{2.10}$$

or

$$\mathrm{W}(M) = P \cdot \mathrm{N}(M) \tag{2.11}$$

Finally, for the evaluation of fractionation data (see Sect. 4.3.2) one needs a *cumulative* (integral) *mass distribution function* to make possible the derivation

**Table 2.1.** Distributions of molecular weight

| Kind of distribution | Value of distribution | Normalization condition |
|---|---|---|
| Frequency distribution (differential) | $N(M)$ | $\int_0^\infty N(M)dM = 1/\bar{P}_n$ |
| | $N(P) = M_o \cdot N(M)$ | |
| Mass distribution (differential) | $W(M)$ | $\int_0^\infty W(M)dM = 1$ |
| | $W(M) = P \cdot N(M)$<br>$W(P) = M_o \cdot W(M)$ | |
| Cumulative distribution (integral) | $I(M) = \int_0^M W(M)dM$<br>$I(P) = I(M)$ | $\int_0^\infty W(M)dM = 1$ |

of the differential mass distribution function $W(M)$ and of the (differential) frequency distribution function $N(M)$.

This distribution function reads

$$I(M) = \int_0^M W(M)dM \qquad (2.12)$$

i.e., $I(M)$ summarizes the macromolecules with MW between 0 and M, related to one gram of polymer. If all components of the polymer sample are summarized (the upper limit of integration goes to infinity), then $I(M) = 1$.

Table 2.1 gives a survey about the different distribution functions. Graphs of distributions will be presented and discussed in Sect. 4.3.2.

## References

1. Elias HG, Bareiss R, Watterson JG (1973) Adv Polym Sci (Fortschritte der Hochpolymeren-Forschung) 11: 111
2. Schröder E, Müller G, Arndt KF (1988) Polymer characterization. Hanser Publ Munich, (1989) Polymer characterization. Akademie-Verlag, Berlin
3. Booth C, Colclough RO (1989) In: Booth C, Price C (eds) Comprehensive polymer science, vol 1. Pergamon, Oxford, p 55
4. Kamide K (1989) In: Booth C, Price C (eds) Comprehensive polymer science, vol 1. Pergamon, Oxford, p 75
5. Huglin MB (ed) (1972) Light scattering from polymer solutions. Academic, London
6. Kratochvil P (1987) Classical light scattering from polymer solutions. Elsevier, Amsterdam (Polymer Science Library, vol 5)

7. Katime IA, Quintana JR (1989) In: Booth C, Price C (eds) Comprehensive polymer science, vol 1. Pergamon, Oxford, p 103
8. Meyerhoff G (1961) Adv Polym Sci (Fortschritte der Hochpolymeren-Forschung) 3: 59
9. Bohdanecký M, Kovař J (1982) Viscosity of polymer solutions. Elsevier, Amsterdam (Polymer Science Library, vol 2)
10. Lovell PA (1989) In: Booth C, Price C (eds) Comprehensive polymer science, vol 1. Pergamon, Oxford, p 173
11. Tung LH, Moore JC (1977) In: Tung LH (ed) Fractionation of synthetic polymers. Marcel Dekker, New York, p 545
12. Glöckner G (1987) Polymer characterization by liquid chromatography. Elsevier, Amsterdam
13. Dawkins JV (1989) In: Booth C, Price C (eds) Comprehensive polymer science, vol 1. Pergamon, Oxford, p 231
14. Tang AQ (1985) Statistical theory of polymer reactions. Academic, Beijing
15. Peebles LH Jr (1971) Molecular weight distribution in polymers. Polymer Reviews (eds: Franke HF, Immergut EH), vol 18. Wiley-Interscience, New York

# 3 Dissolution and Precipitation of Polymers

## 3.1 Rules for Solubility of Polymers

Solubility of a polymer in a solvent, which is the most important property for polymer fractionation, depends generally on the chemical structures of both polymer and solvent. The old rule "*similia similibis solventur*" (solubility results from similarity) may give useful hints to select solvents for a polymer. This can be seen for the most solvent/polymer pairs listed in Table A 3 in the Appendix. Sometimes however, this rule also fails and this is illustrated by the following example: Water is a good solvent for poly(vinyl alcohol); poly(vinyl acetate) (from which poly(vinyl alcohol) usually is derived) is soluble in ketones. However, methanol or ethanol are nonsolvents for poly(vinyl alcohol) but solvents for poly(vinyl acetate).

Solubility decreases with increasing MW. A rise in temperature usually enhances the solubility, but, for a system with low critical solution temperature (LCST – see Glossary; one of a few examples is the system poly(vinyl alcohol)/water), the opposite holds good.

The solubility of copolymers depends mainly on CC and is generally better than expected from the solubilities of the parent homopolymers. For instance, both poly(styrene) and poly(acrylonitrile) are insoluble in acetone, but, a styrene/acrylonitrile copolymer with middle composition is soluble.

The structure of a polymer solution can depend on concentration. Often, molecular-disperse solutions exist only in a sufficiently diluted state whereas clusters (associates, aggregates) may be formed with increasing concentration. This is possible especially in the vicinity of the theta state (see Glossary) where interactions between the macromolecules become operative. Therefore, fractionations should be carried out at low concentrations.

Solubility problems arising from secondary structures in the polymer sample (supermolecular or ordered structures) have already been mentioned in Sect. 1.1. Ordered structures, especially crystallinity of a polymer, impede dissolution. Usually, the dissolution of a crystalline polymer only succeeds when the polymer/solvent system is heated to above the melting point of the crystallites.

The driving force for the solubility of a polymer in a solvent is provided by the difference in the chemical potentials of the two species. This reads in terms of the Gibbs-Helmholtz equation

$$\Delta G_{mix} = \Delta H_{mix} - T\Delta S_{mix} \tag{3.1}$$

where $\Delta G_{mix} < 0$ is the free energy of mixing. The entropy of mixing, $\Delta S_{mix} > 0$, is large for low-MW substances but much smaller for polymers due to the size of polymer molecules and the repeat units maintaining their "linear-ordered structure" in solution. Thus, the dissolution of polymers requires a rather small value of the enthalpy of mixing. This is the reason why the solubility of polymers is much more restricted than that of respective low-MW compounds.

Let us turn to concepts for the quantitative description of the solubility of polymers.

**Solubility parameters:**

The enthalpy of mixing is according to Hildebrand [1] proportional to the square of the difference in the so-called solubility parameters for the solvent (S), $\delta_S$, and the polymer (P), $\delta_P$,

$$\Delta H_{mix} \propto (\delta_S - \delta_P)^2 \tag{3.2}$$

with

$$\delta_S = (\Delta E_S/V_S)^{0.5} \tag{3.3a}$$

and

$$\delta_P = (\Delta E_P/V_P)^{0.5} \tag{3.3b}$$

$\Delta E_S$ and $\Delta E_P$ are the values of the internal heat of evaporation, $V_S$ and $V_P$ are the molar volumes of solvent and polymer, respectively. Thus, solubility parameters represent the interactions in pure substances which must be overcome in the dissolution process.

Solubility parameters of solvents can be measured directly whereas for polymers, they must be determined in indirect ways. Estimation via additivity of increments for different structural units is also possible (see, for instance, [2] and Table A 1 in the Appendix). For copolymers and mixtures, the solubility parameters can be calculated assuming additivity of the respective components using volume fractions $\varphi$:

$$\delta = \varphi_A\delta_A + \varphi_B\delta_B \tag{3.4}$$

where indices A and B represent parent homopolymers or two solvents (or solvent and nonsolvent), respectively.

As can be seen from Eq. (3.2), the enthalpy of mixing is a positive quantity. Its lowest value, which characterizes the best solubility, occurs for $\delta_S = \delta_P$. The solubility parameters of common solvents and polymers have been tabulated (see, for instance, [2]). They are a useful tool for the estimation of the solubility in advance. Solubility parameters of selected solvents and nonsolvents are given in Table A 2 in the Appendix.

Table A 3 summarizes solvents and nonsolvents for polymers ordered according to their solubility parameters. As can be seen, most solvent/polymer pairs yield differences in their solubility parameters of less than two whereas these differences are mostly larger for nonsolvent/polymer pairs. Deviations from these rough rules occur sometimes with polar substances.

According to Eq. (3.4), use of solvent/nonsolvent mixtures enables us to vary the difference in the solubility parameters of (mixed) solvent and polymer. This possibility is extensively utilized in fractionation procedures.

**Polymer/solvent interaction parameters (Huggins constants):**

The Huggins constant $\chi$ in the Flory-Huggins equation [3, 4]

$$\Delta\mu_S/RT = \ln(1 - \varphi_P) + (1 - 1/P)\varphi_P + \chi\varphi_P^2 \tag{3.5}$$

is also a measure of solubility. In this equation, $\Delta\mu_s$ is the chemical potential of the solvent in solution, $\varphi_P$ is the volume fraction of the dissolved polymer having a DP of $P$ and R and $T$ are gas constant and temperature, respectively.

For a polymer to be soluble it is necessary for $\Delta\mu_S \leqslant 0$. Here we have to mention the very important aspect of phase stability of polymer solutions. Like any other solution, a polymer solution can also decay under certain conditions into a heterogeneous system consisting of two phases. These phases are characterized by different compositions. As long as $\Delta\mu_S$ behaves as a monotonously decreasing function of volume fraction $\varphi_P$, the solution stays homogeneous. However, when a point of inflection appears, then phase instability occurs (cf. Fig. 3.1). In mathematical terms, the critical conditions are: $d\Delta\mu_S/d\varphi_P = 0$ and $d^2\Delta\mu_S/d^2\varphi_P = 0$. For the critical value of $\chi$ it follows that

$$\chi_{crit} = [1 + (1/P)^{0.5}]^2/2 \tag{3.6}$$

which results in $\chi_{crit} = 0.5$ for $P \to \infty$.

In Fig. 3.1 one can distinguish curves with and without extrema depending on DP. In other words, values of $\chi > \chi_{crit}$ lead to phase separation into a polymer-rich gel phase and a highly diluted sol phase (see dashed curve for $\chi = 0.625$). Curves without extrema occur for $\chi < \chi_{crit}$. Under these conditions, the system is homogeneous and the polymer miscible with the solvent. Figure 3.1 shows that the critical curve depends on DP (see, e.g., the curves for $\chi = 0.5$ ($P = \infty$) and $\chi = 0.605$ ($P = 100$)) which results from Eq. (3.6). $\chi_{crit}$ increases with decreasing DP, i.e., phase stability is more extended.

The Huggins constant can be related to the solubility parameters:

$$\chi = (V_S/RT)(\delta_S - \delta_P)^2 \tag{3.7}$$

For a given temperature $T$, $\chi$ depends on both the difference in the solubility

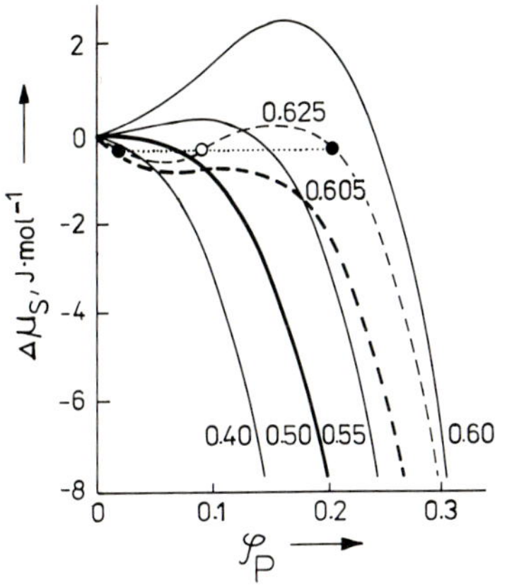


**Fig. 3.1.** Dependence of chemical potential of the solvent on volume fraction of the dissolved polymer according to Eq. (3.5). *Full lines* for $P = \infty$ and $T = 298.2$ K; *dashed lines* for $P = 100$ and $T = 298.2$ K; *bold curves* represent the critical state for corresponding DP. The *dotted line* shows phase separation in a sol phase (*left point*) and a gel phase (*right point*) characterized by equal $\Delta\mu$s values. The *numbers* indicate values of $\chi$. (Adopted from Ref. [5] with permission of Hüthig Verlagsgemeinschaft GmbH, Deutscher Verlag der Wissenschaften, Berlin)

parameters of solvent and polymer and the molar volume of the solvent. For $\delta_S = \delta_P$, it gives $\chi = 0$.

The value $\chi_{crit}$ is attained for a certain difference in solubility parameters dependent on temperature and solvent ($V_S$). In a solvent/nonsolvent mixture, the value $\delta_S$ can, according to Eq. (3.4), be adjusted to the critical $\chi$ value. Consequently, phase separation can be initiated. From Eq. (3.7), the temperature dependence of the phase stability can be also derived. For a given system, the Huggins constant decreases with increasing temperature, i.e., a system unstable at low temperatures can become stable upon heating. This effect can be understood also on the base of Eq. (3.1) where rising temperature increases the product $T \cdot \Delta S$.

$\chi = 1/2$ characterizes the so-called theta state (see Glossary). According to Eq. (3.7), this state is fixed for a certain polymer/solvent system by a characteristic temperature, the theta temperature. In mixed solvents, the theta temperature is related to a particular ratio of the components, the theta composition of the solvent. When the temperature in a polymer/(mixed) solvent system is reduced below the theta temperature, then, we notice the appearance of separation into sol and gel phase.

Values of polymer/solvent interaction parameters are summarized in Ref. [6].

## 3.2 Dissolution and Precipitation Equilibrium

According to Eq. (3.6), phase separation into the concentrated gel phase and the diluted sol phase occurs when the interaction parameter $\chi$ exceeds its critical value. As a starting-point, one can use either a diluted solution of the sample or, directly, the polymer to be investigated. In the former case, the gel fraction is obtained by precipitation, while in the latter the sol fraction is extracted. In order to achieve fast adjustment of the equilibrium, the polymer should be applied as a very thin layer, e.g., as a coating on an unswollen support material. Since the polymer film swells, the equilibrium is also in this procedure the coexistence of sol and gel phase.

Description of these equilibria is possible by phase diagrams [5, 7–11]. Let us discuss the simplest case – a system consisting of a solvent S and two polymer-homologous species $P_1$ and $P_2$ where $P_1$ represents the smaller polymer molecules. We will adopt an example given by Koningsveld et al. [11]. The conclusions are also valid – at least qualitatively – for more complicated systems as a polymolecular polymer in a solvent/nonsolvent mixture.

Figure 3.2 shows the solvent-rich corner of the phase diagram for solutions of two polymer-homologues $P_1$ and $P_2$ in a single solvent S. Since $P_1$ and $P_2$ have the same chemical composition, interaction parameters $\chi_{SP1}$ and $\chi_{SP2}$ are

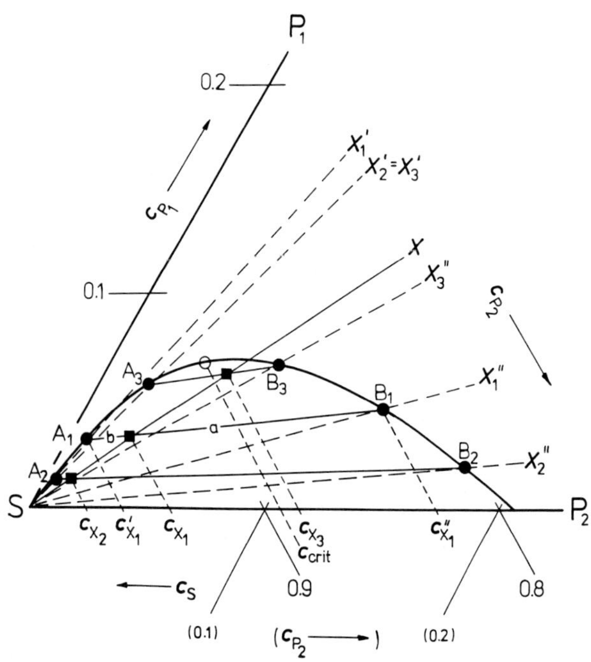


**Fig. 3.2.** Ternary phase diagram for solutions of two polymer-homologues $P_1$ and $P_2$ in a solvent $S$, calculated for $\chi_{SP1} = \chi_{SP2} = \chi = 0.5988$. Concentrations $c_{P_2}$ are projected to the $\overline{SP_2}$ line of the triangle (*written in parentheses*). Lines marked with $X$, $X_i'$ and $X_i''$ represent solutions or phases having constant $P_1/P_2$ ratios. The *solid curve* encloses the two-phase region. ○ – critical point; ●—● – tie line of coexisting phases; $A_i$ and $B_i$ – cloud points for binary polymer mixtures $X_i'$ (sol phase) and $X_i''$ (gel phase), respectively. (Reprinted from Ref. [10], Comprehensive Polymer Science, ed. by Allen, vol. 1 (eds. Booth, Price), p. 293, Copyright 1989, with kind permission from Pergamon Press Ltd., Headington Hill Hall, Oxford OX3 OBW, UK)

assumed to be identical ($\chi_{SP1} = \chi_{SP2} = \chi$). Lines marked with $X$, $X'_i$ and $X''_i$ represent polymer mixtures having constant proportions of $P_1$ and $P_2$. (Single primed quantities correspond to sol phases, double primed ones to gel phases). Thus, all points on line $\overline{XS}$ correspond to the same MW average but to different overall polymer concentrations $c_{x_i}$. The greater the distance from S, the higher the concentration ($c_{x_3} > c_{x_1} > c_{x_2}$). Values $c_{x_i}$, $c'_{x_i}$ and $c''_{x_i}$ of the total polymer concentration are projected to the $\overline{SP_2}$ line corresponding to $c_{x_i} = 1 - c_S$. The solid curve surrounds the two-phase region. The extension of the latter depends on the $\chi$ value. A polymer solution $X$ with the concentration $c_{x_1}$ inside the two-phase region will separate into the sol phase $A_1$ with concentration $c'_{x_1}$ and polymer composition $X''_1$ and the gel phase $B_1$ with concentration $c''_{x_1}$ and polymer composition $X''_1$, respectively. The result is a fractionation effect leading to different compositions $X'_1$ and $X''_1$ for the sol and gel phase, respectively. The ratio of the axis a and b on the tie line (which connects the coexisting phases A and B) represents the volume ratio of the phases A and B, respectively. The same can be achieved with a smaller concentration $c_{x_2}$ and a higher one, $c_{x_3}$. Comparison of the results for the different concentrations leads to the following conclusions important for the practice of fractionation:

(i) Separation of the pure components $P_1$ and $P_2$ is impossible.
(ii) Lowering of the initial polymer concentration improves considerably the fractionation efficiency for the gel phase (the amount of $P_1$ in this fraction decreases) whereas the polymer composition in the sol phase, $X'$, alters only slightly.
(iii) The volume ratio $V''/V'$ ($=$ b/a) decreases with diminution of the concentration. That means that the improvement of the fractionation efficiency for the gel phase is connected with a smaller amount of polymer in this fraction. With a concentration $c_{x_3}$ above the critical concentration $c_{crit}$ (see Glossary – Critical Point), it results $V''/V' > 1$. Now, the fractionation efficiency is better for the sol phase than for the gel phase. This is the case of extraction fractionation where higher polymer concentrations can be used without deterioration of fractionation efficiency.

If the temperature of the system is lowered, the two-phase region grows in such a manner that the extension is relatively small against line $\overline{SP_1}$ but large in the direction of line $\overline{SP_2}$. The fractionation efficiency improves for the sol fraction but not for the gel fraction which contains more of the polymer. In other words, if one precipitates an extended gel phase (say, 90% of the initial polymer), one obtains a sol phase with composition $X'$ near to the $\overline{SP_1}$ line. The gel phase can be re-dissolved and the precipitation can be repeated. After some repetitions, this is a precipitation fractionation suggested by Staverman and Overbeek (see [9]). Here, the first fraction (what is a sol fraction) contains the molecules having lowest MWs and for the next fractions MWs go up.

Therefore, this special kind of precipitation fractionation shall be called "*upward precipitation fractionation*" in this book , in contrast to the "normal" precipitation fractionation where MWs go down with increasing fraction number. (The latter procedure could be named "downward precipitation fractionation".) Since the concentration dependence of the fractionation efficiency for the sol phase is weak in "upward precipitation fractionation", one can proceed to fractionate a polymer at higher concentration.

In the following sections, the short expression precipitation fractionation refers to "downward precipitation fractionation".

Conclusions drawn from Fig. 3.2 can be extended qualitatively to systems having a continuous MWD. If the precipitation is caused by addition of a nonsolvent instead of temperature decrease, the consequences for fractionation seen from a practical point of view are very similar.

For the quantitative description of the phase conditions, the distribution coefficient $K_i$ has been introduced [7]. $K_i$ represents the ratio of the polymer concentrations $c_i''$ and $c_i'$ in the gel and sol phase, respectively. It depends on DP:

$$K_i = c_i''/c_i' = \exp(k \cdot P_i) \tag{3.8}$$

The quantity $k$ is positive and gets bigger with increasing $\chi$, i.e., with advanced precipitation. Substitution of $c$ gives:

$$w_i''/w_i' = (V''/V')\exp(k \cdot P_i) \tag{3.9}$$

Rearrangement gives the mass $w_i''$ of molecules with $P_i$ in the gel phase,

$$w_i'' = w_i/[1 + V'/V'')\exp(-k \cdot P_i)] \tag{3.10}$$

The mass $w_i'$ of molecules with $P_i$ in the sol phase follows from $w_i' + w_i'' = w_i$ and after rearrangement as

$$w_i' = w_i/[1 + (V''/V')\exp(k \cdot P_i)] \tag{3.11}$$

$w_i$ is the mass of molecules with $P_i$ in the unfractionated polymer, $V'$ and $V''$ are the volumes of the corresponding phases. These equations were first derived by Schulz [7] starting from experimental results. They are also available from Flory-Huggins theory [4, 9, 11].

It follows from Eqs. (3.9) and (3.10) that fractionation efficiency, which depends upon $P_i$, is diminished for the gel phase by a higher overall polymer concentration as well as by a higher $\chi$ value (i.e., by a greater amount of precipitated polymer). Thus, the gel phase is contaminated with more low-molecular constituents than in the case of lower concentration and lower $\chi$ values. It is possible to ascertain these influences by determining the "non-uniformity" $U$ of the gel fraction, e.g., by SEC.

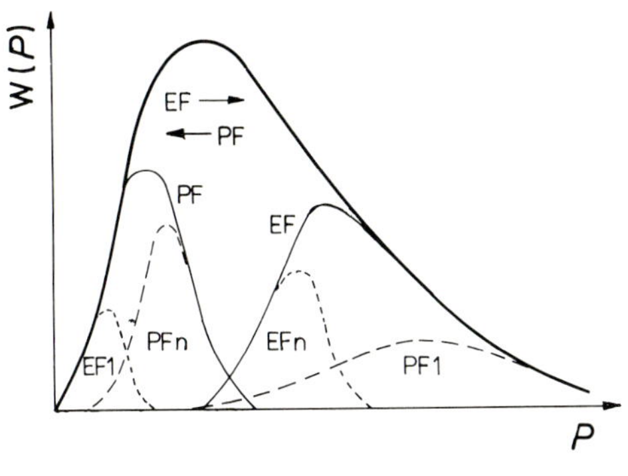


**Fig. 3.3.** Comparison of precipitation fractionation (*PF*) and extraction fractionation (*EF*) (schematically). —— initial distribution, —— residual distributions after n − 1 fractionation steps. Directions of fractionation are indicated at the top

Unlike the gel phase, fractionation efficiency for the sol phase is less affected by concentration. It improves continuously with increasing gel phase. These effects are understandable when one considers the general characteristics of the mass distribution $W(P)$ illustrated in Fig. 3.3. The bold solid curve reflects the initial mass distribution of a polymer. If one performs a precipitation fractionation (PF), the gel fraction PF1 is obtained first. According to Eq. (3.10), $w_i''/w_i$ decreases with descending $P_i$. If, however, $w_i$ (or $W(P)$) grows with decreasing $P_i$, $w_i''$ can increase with decreasing $P_i$. This is the reason for the "low-molecular tailing" which is a characteristic and inherent feature especially of fractions obtained from precipitation fractionation (see also Sect. 4.3.2 and Fig. 4.2). Consequently, fractionation efficiency worsens, especially with increasing polymer concentration. In a later precipitation step which yields fraction PFn, one finds very similar conditions owing to the further increase of $W(P)$ with decreasing $P_i$ (cf. the fine solid curve denoted by PF).

For extraction fractionation (EF), the problems are the opposite. Here one obtains at first the sol fraction EF1 with increasing $w_i$ for growing $P_i$, i.e., the effect of Eq. (3.10) is intensified. In Eq. (3.11), both numerator and denominator increase in this case and $w_i'$ is influenced only a little. Conditions remain unchanged in later extraction steps (see fine solid curve EF and sol fraction EFn).

These considerations lead us to the conclusion that extraction fractionation is more efficient than precipitation fractionation. "Upward precipitation fractionation", however, must be discussed in analogy to extraction fractionation because the first fraction would be like EF1.

Finally, a phenomenon that sometimes appears in fractionation – the so-called "reverse-order precipitation" [7] or, more generally, "*reverse-order fractionation*" – is related to influences discussed above. It comes about that when one uses higher polymer concentrations and steep changes in $\chi$ (for instance, by use of a very strong precipitant). The sequence of fractions is then not strictly ordered with respect to MW, i.e., inversions occur.

## References

1. Hildebrand J, Scott R (1949) The solubility of nonelectrolytes, 3rd edn. Reinhold, New York
2. Grulke EA (1989) In: Brandrup J, Immergut EH (eds) Polymer handbook, 3rd edn, Wiley, New York, p VII/519
3. Kamide K (1989) In: Booth C, Price C (eds) Comprehensive polymer science, vol 1. Pergamon, Oxford, p 75
4. Huggins ML, Okamoto H (1967) In: Cantow MJR (ed) Polymer fractionation. Academic, New York, p 1
5. Glöckner G (1987) Polymer characterization by liquid chromatography. Elsevier, Amsterdam
6. Gundert F, Wolf BA (1989) In: Brandrup J, Immergut EH (eds) Polymer handbook, 3rd edn. Wiley, New York, p VII/173
7. Schulz GV (1940) Z Phys Chem B 46: 137; (1953) In: Stuart HA Die Physik der Hochpolymeren, vol II. Springer, Berlin, Göttingen, Heidelberg, p 726 (Chapter 17)
8. Tompa H (1956) Polymer solutions. Butterworth, London
9. Koningsveld R (1970) Adv Polym Sci 7: 1
10. Casassa EF (1977) In: Tung LH (ed) Fractionation of synthetic polymers. Marcel Dekker, New York, p 1
11. Koningsveld R, Kleintjens LA, Geerissen H, Schützeichel P, Wolf BA (1989) In: Booth C, Price C (eds) Comprehensive polymer science, vol 1. Pergamon, Oxford, p 293

# 4 Basic Principles of Fractionation Procedures

## 4.1 Preparations for a Fractionation

A fractionation procedure can be planned more efficiently when the knowledge about the polymer is more exact. Therefore, the importance of the preparation for a fractionation must not be underestimated. Such a preparation involves

- information about characteristics and properties of the *polymer*
- knowledge about properties of *solvents and nonsolvents* (related to polymer and fractionation procedure)
- choice of the *fractionation technique* in consideration of aim and expenditure of the fractionation.

### 4.1.1 Polymer

Questions concerning the polymer are summarized in Table 4.1 with added hints about possible answers. The succession of these questions corresponds to their importance for fractionation.

Usually, information on the *chemical structure* of the polymer should be provided by the supplier. Otherwise, one has to use qualitative and quantitative analyses. Of course, the variety of methods is broad and the method of choice depends on the polymer. Some guides in the literature [1–7] can help to solve the problem (see Table 4.1). If nothing is known about the polymer, some simple identification tests can yield qualitative information about the kind of polymer. Table A 4 in the Appendix contains simple characteristics of some soluble polymers for identification. More detailed qualitative and quantitative analyses must be performed according to Table 4.1.

Besides, the following questions should be checked in relation to the chemical structure of the polymer:

- *Is it necessary to stabilize the polymer solution*? Fractionations can be very time consuming and may require elevated temperatures. Therefore, it may be necessary to add a suitable stabilizer – for instance against oxidation – to the solution.

**Table 4.1.** Preliminary investigations of the polymer assigned for fractionation

| Questions | Possible methods for answer |
|---|---|
| – Chemical structure? (monomer units) → stabilization? → chemical reactions with solvent or precipitant? | Qualitative analysis [1–5, 7], solubility [6], appearance, spectroscopy [5], pyrolysis (coupled with mass spectrometry or gas chromatography) [5], quantitative chemical analysis (especially for co-polymers) |
| – Solubility? (completeness, crystalline or crosslinked residues) | Dissolution tests [6] |
| – Additives? (catalysts, inhibitors, emulsifiers, stabilizers, fillers) | Spectroscopy [5] |
| – Range of molecular weight? | Osmotic pressure or end-group analysis ($\bar{M}_n$) [8], light scattering ($\bar{M}_w$) [9], viscometry ($\bar{M}_v$) [10], SEC [11], see also textbooks of physical chemistry of polymers) |
| – Conditions of manufacture? • mechanism • temperature of reaction • conversion • monomer ratio (copolymers) • continuous/discontinuous reaction (especially for copolymers) • isolation of polymer (precipitation or not) | Polymer synthesis |
| • drying (temperature, vacuum, intentensity) | Weight checking after some repetitions of drying |

– *Do chemical reactions of the polymer* with solvents or precipitants occur? Sometimes, polymers or copolymers contain reactive monomer units (e.g., anhydride units). Solvents and nonsolvents capable of chemical reactions with the polymer sample should be avoided (the long contact time at elevated temperatures during a fractionation should be considered).

*Solubility of the polymer sample* in the solvent is the fundamental point of fractionation. A first simple test can be performed by putting a solvent droplet on the polymer (powder or surface). After sufficient time, the polymer is sticky if a solvent was used. The polymer must be completely soluble to obtain a good fractionation. Complete solubility can often be achieved via change of solvent or heating. For complete dissolution, some hours may be necessary. It should be checked that the polymer is not just swollen and invisible in the solvent due to similar refractive indices. Nevertheless, solutions may comprise insoluble (cross-linked or crystalline) portions. For crystalline structures, one has to heat beyond the melting point of the crystallites. When the polymer tends to re-crystallize upon cooling, the complete fractionation must be carried out at sufficiently high temperatures. Evidently, crosslinked portions remain insoluble and must be separated from the solution by sedimentation, centrifugation or filtration before the fractionation starts. For solubility tests, one has to take into consideration conditions of fractionation (concentration range, temperature range). In Table A 3 in the Appendix, typical solvents and nonsolvents for some polymers are summarized.

Polymers may contain *additives* as a result of synthesis (catalysts, inhibitors, emulsifiers) or to ensure processing and application (stabilizers, fillers). If one has no information about additives in the polymer sample, spectroscopic investigations [5] are often helpful. Such additives can be removed by dissolution and subsequent re-precipitation (see p. 25) when additives are soluble in the nonsolvent for the polymer. In case of doubt, re-precipitation is generally recommended.

Information about *conditions of both polymer synthesis and polymer isolation* are not necessary for the fractionation but may be helpful. Mechanism of synthesis, reaction temperature, and conversion can influence the structure of the sample (tacticity, branching, crosslinking – see Sect. 1.1). This also provides hints about additives.

Conditions of isolation of the sample from the reaction mixture can influence properties of the polymer. For instance, if the isolation was performed by drying instead of precipitation, all additives will remain in the polymer sample. Temperature, intensity of vacuum, and time of drying belong to the history of the sample and, consequently, influence the sample (ordering processes, ageing, partial degradation, degree of drying; cf. Sect. 1.1). Finally, knowledge about the eventual *range of MW* (width of MWD) is useful and can be obtained advantageously by SEC.

### 4.1.2 Solvent and Precipitant

Solvents and precipitants for fractionation should be selected according to the factors summarized in Table 4.2.

**Table 4.2.** Preliminary investigations on applicable solvents and precipitants

<table>
<tr><td>Solubility strength (solvent)</td><td rowspan="3">Turbidimetric titration (see Sect. 5.4.2)</td></tr>
<tr><td>– Precipitation strength (precipitant)</td></tr>
<tr><td>– Miscibility</td></tr>
<tr><td colspan="2">– Physical characteristics (boiling point, density)</td></tr>
<tr><td colspan="2">– Chemical characteristics (stability, nontoxicity, inflammability)</td></tr>
</table>

Both *dissolution and precipitation power* should be relatively weak. Otherwise, the development of phase equilibria and the sharpness of separation are hampered. Estimate of solvents and nonsolvents can be effected by comparison of the solubility parameters of polymer, solvent and nonsolvent. For instance, using Table A 3, cyclohexane (above 35 °C) should be a more favourable solvent for the fractionation of poly(styrene) than toluene because of the higher difference in the $\delta$ values of poly(styrene) and cyclohexane (cf. Eq. (3.7)). Ethanol is to be preferred as precipitant instead of methanol because of the smaller $\delta$ difference of the former to poly(styrene). Comparison of cloud points (see Glossary) resulting from different precipitants (added to respective polymer solutions in different solvents) may also help us to estimate solvents and nonsolvents and this will be illustrated with the following example. Cloud points of a 50/50 (mole/mole) styrene/acrylonitrile copolymer were determined using dichloromethane and acetone as solvents and methanol and $n$-hexane as precipitants. It was found (as volume fraction of precipitant, $\varphi^*$, at 20 °C):

| | $\varphi^*$ | |
|---|---|---|
| | methanol | $n$-hexane |
| dichloromethane | 0.449 | 0.342 |
| acetone | 0.393 | 0.283 |

Thus, acetone is the poorer solvent and methanol the weaker precipitant. Special problems with the choice of solvent/nonsolvent combinations in the fractionation of copolymers will be discussed in Sect. 4.4. Mixtures of solvents and nonsolvents can be used also as solvents or nonsolvents to adjust to suitable conditions. Such investigations can be performed advantageously by turbidimetric titration (see Sect. 5.4.2). *Miscibility* (absence of miscibility gaps) of solvent and precipitant can be checked also with this method.

For drying of polymer fractions, low *boiling points* of solvents and nonsolvents are desirable. However, dissolution and fractionation temperatures must be taken into account.

The rate of phase separation, e.g., in precipitation fractionation, depends on *densities of solvent and precipitant*. Separation of sol and gel phases proceeds faster when the difference in their densities is higher. This difference depends on the composition of both phases. Generally, the volume fraction of the solvent is higher in the gel phase than in the sol phase. Indeed, the polymer concentration in the gel phase is higher than that in the sol phase, but, it does not exceed a few percent of the gel phase. Thus, the solvent is the main component in the gel phase. Fortunately, the most common precipitants are lower alcohols and hydrocarbons having low densities and that is why the gel phase usually has the higher density. Of course, the polymer density also plays a role. In a gradient fractionation (see Sect. 7), problems with the generation of gradients may arise when densities of solvents and nonsolvents are too disparate. Similar problems concerning phase separation and densities of solvents occur in partition fractionation (see Sect. 11).

*Chemical stability* concerns not only the polymer to be fractionated but also solvent and precipitant. (For instance, THF and other ethers can form dangerous peroxides.) Unstable species should be avoided or must be stabilized.

For a choice of solvents and nonsolvents, one should consider its potential *toxicity*. If it is not possible to replace a toxic solvent or precipitant by a nontoxic substance, the fractionation must be carried out with special care. An exhauster is in general the recommended place for performing a fractionation if arrangement, construction, and dimensions of the apparatus allow this.

Many organic solvents used in fractionation procedures are *flammable*. Therefore, electric and heated devices should be arranged carefully to prevent any danger caused by breakage of parts containing combustible liquids.

In practice, it is sometimes impossible to satisfy simultaneously all these requirements. Thus, a fractionation is often subjected to expedient compromises. The best preparation for a successful fractionation is a test run using only a small amount of polymer.

### 4.1.3 Fractionation Technique

The choice of a particular fractionation technique depends on the aim of the fractionation. Additionally, conditions and possibilities of the laboratory and economy may play a role. Let us discuss some general aspects of this problem (details are considered in the respective sections describing individual procedures).

The determination of MWD of a polymer is still seldom the aim of a fractionation. It would be possible, on principle, by analytical fractionation, but, SEC is the usual method nowadays.

Determination of MW dependence on a certain polymer property requires large enough fractions obtainable via fractionation on a preparative scale. However, the amount of one fraction should not exceed ten percent of the total mass of polymer. Otherwise, the fractionation efficiency (sharpness of fractions) is not sufficient.

Concerning the expense of different techniques, one may establish roughly the following succession: extraction fractionation without a gradient <precipitation fractionation (<“upward precipitation fractionation”) <gradient elution < elution using two gradients (solvent and temperature). Additionally, some other aspects have to be recognized. Precipitation fractionation is the most time-consuming method, but, it needs much less preliminary investigation of the fractionation in comparison with, say, gradient elution. Fractionation by partition between immiscible liquids (see Sect. 11) can be performed very easily but, the preparatory investigations may be extensive.

Sometimes, the consumption of solvents and nonsolvents governs the choice of the method used. Roughly, precipitation fractionation usually requires less solvent and precipitant than "upward precipitation fractionation" or gradient elution. However, these quantities must be related to the mass of polymer which can be fractionated in one run (regulated by the possible polymer concentration).

## 4.2 General Rules for Implementation of Fractionations

### 4.2.1 Sample Preparation and Phase Equilibrium

A fractionation usually starts with a solution. This is obvious for a precipitation fractionation. In an extraction (elution) fractionation, the formation of the gel phase as a thin layer also starts from a polymer solution. Generally, polymers are less stable in solution than in the solid state. Light (especially the UV component), heat, and chemical influences may attack polymer molecules in solution. Therefore, polymer solutions should be fractionated or precipitated immediately after the time necessary for complete dissolution. The fractionation should be performed steadily, i.e., without needless breaks in fractionation. Sometimes, it is necessary to protect the polymer solution against light. Before one dissolves the polymer, two aspects should be considered:

- Is it necessary to re-precipitate the polymer sample to eliminate additives (cf. p. 21)? Details of re-precipitation will be discussed in the next section.
- Was the polymer sample dried sufficiently? Residues of solvents or humidity in the polymer can affect the course of fractionation. Some hints to the drying process will be given below in connection with the isolation of fractions (p. 26).

Another general problem in a fractionation is the quick and complete *adjustment of the equilibrium between sol and gel phases* which requires an extended interface between the phases. This is ensured by stirring in precipitation and partition fractionation and by sufficiently thin gel layers in extraction and elution fractionation.

A second aspect related to the equilibrium is the rate of changing precipitation or dissolution power in the course of fractionation which should be slow. Minor variation in precipitation or dissolution power results from use of poor solvents and nonsolvents (as discussed in Sect. 4.1.2) and can be assisted by slow addition of the precipitant or by use of flat gradients.

Finally, let us supplement *practical aspects for fractionation* derived from discussion of phase equilibria (Sect. 3.2):

(i) Low polymer concentrations as well as small volume ratios $V''/V'$ improve fractionation efficiency (especially in precipitation fractionation).
(ii) Growing masses of fractions usually cause an increase of the non-uniformity of the fractions.
(iii) Sol phases are less affected by concentration, MWD, and masses of fractions than gel phases. Thus, "upward precipitation fractionation" and extraction (elution) fractionation should be more efficient.
(iv) Danger of "reverse-order fractionation" increases with increasing polymer concentration and rapid changes of dissolution or precipitation power of the system.

### 4.2.2 Re-dissolution and Re-precipitation of Fractions

Each fractionation yields fractions either as a sol phase (extraction, elution) or as a gel phase (precipitation). Gel phases must be re-dissolved and afterwards re-precipitated to enhance their sharpness. For re-dissolution and re-precipitation, there are some generally accepted rules independent of the fractionation technique used:

*Re-dissolution* of the isolated gel phase should be performed with a small amount of the same solvent used in the fractionation procedure. It must be taken into consideration, however, that very powerful and high-boiling solvents can only be removed from the polymer with great difficulty. Then, the solvent should be changed, but, one has to ascertain that the isolated fraction is *completely* soluble in the chosen solvent.

*Re-precipitation* should ensue in a strong, but easily volatile nonsolvent at concentrations of the polymer solution of about one percent. *Vigorous agitation* is necessary during and for a time after *dropwise addition* of the dissolved gel phase to the precipitant. Volume ratio solution/precipitant should be at least 1:10 (1:20 is better). A low precipitation temperature is usually chosen to obtain a good flocculation and to prevent stickiness of the precipitate (on condition that LCST behaviour does not exist; see Glossary). Agitation after finishing of re-precipitation is necessary for the growth of the polymer particles to give a freely filterable precipitate. For re-precipitation, a precipitant should never be added to the solution because of the possibility of fractional precipitation of the polymer.

After sedimentation, if necessary after centrifugation, the supernatant should be decanted and fresh precipitant must be added to the precipitated polymer owing to a further exchange of solvent and nonsolvent in the precipitate. (A nonsolvent can be removed more easily than a solvent. A solvent-free polymer does not stick in further handling.) Finally, it should be noted that the consistency of a precipitated fraction varies considerably (solid, stringy, glassy, powdery, . . . ). The precipitate must be filtered from the supernatant after re-precipitation, solvent/nonsolvent exchange, and sedimentation. If the polymer is not sensitive to oxidation, the sample should be dried a short time on air (after filtration) to remove adhered nonsolvent.

### 4.2.3 Drying

Drying is the last step in the isolation of fractions. The filtered fraction must be dried in *reasonable vacuum*, i.e., about 130 Pa or less. It is decisive that one does not switch off the vacuum pump or, at least, that it works with only short breaks. Otherwise, an equilibrium develops in the closed system without progress in drying. Elevated temperatures are desirable, if possible (with due regard to degradation reactions at higher temperatures) higher than the boiling temperature of the solvent and nonsolvent in vacuum and the glass-transition temperature of the polymer (see Glossary). The drying process must be continued until *weight constancy* of the dried polymer is approached (some days may be necessary).

It should be mentioned that samples having different degrees of dryness must not be dried together. Otherwise, the lowest-dry sample determines the period of time necessary for drying of all other samples.

## 4.3 Algorithm for Evaluation of Fractionation Data

Results of fractionation procedures are masses or mass fractions of the individual polymer fractions. For evaluation of the fractionation, in addition, one needs quantities characterizing these fractions. Usually, the MWD is obtained via SEC or, at least, one of the MW averages is measured with an appropriate method (see Sect. 2.1 and Table 4.1). Sometimes, one must be satisfied (when direct measurements of MW are lacking) with equivalent quantities such as SEC elution volume $V_e$, intrinsic viscosity $[\eta]$, or cloud points $\varphi^*$ (expressed by the volume fraction of precipitant in turbidimetric titration, see Sect. 5.4.2). Please, note that these equivalent values are in a constant relation to MW only if all fractions have the same molecular structure (see Glossary, KMHS equation). When a copolymer was fractionated, the CC of each fraction should be determined additionally (see Sect. 4.4). Thus, we obtain as *basic data for each fraction*:

(i) mass or mass fraction,
(ii) MWD or a quantity describing the mean size of molecules and, in the case of copolymers,
(iii) CC (which can also be an average).

These results should be summarized in a table, an example of which is Table 4.3.

### 4.3.1 Judgement of Fractionation Quality and Efficiency

There are *criteria for fractionation quality* which should be checked after a fractionation using the quantities mentioned above:

(i) *Mass balance*: After drying of all fractions, the sum of their masses should be equal to the mass of the dissolved polymer. Usually, the mass of all fractions is somewhat lower (a few percent only). Sometimes, the total mass of fractions may exceed that of the starting polymer. Reasons and consequences of that will be discussed below (Sect. 4.3.2). If one finds mass differences smaller than three or four percent, the result is acceptable. An additionally important point of view is the relative mass of the individual fractions which usually should be nearly equal.

(ii) *Averages of MW*: If number, weight, or viscosity averages of MW have been measured for the unfractionated polymer, it is possible to compare these directly measured values with corresponding values calculated from masses and MW of all fractions. These calculations can be performed according to Eqs. (2.2) through (2.4). Of course, correspondence can only be expected when calculated and measured values are averaged by the same algorithm, e.g., number average $\bar{M}_n$ calculated from the $\bar{M}_n$ values of fractions and measured $\bar{M}_n$ of the unfractionated sample, etc. SEC renders the determination and comparison of all MW averages.

(iii) *Average of intrinsic viscosity*: Intrinsic viscosity is a weight average of the values for the individual components,

$$[\eta] = \sum (w_i\,[\eta]_i)/\sum w_i \tag{4.1}$$

Thus, we can compare measured values $[\eta]$ of starting polymers and calculated values according to Eq. (4.1) using the measured quantities $[\eta]_i$ and $w_i$ of all fractions.

These criteria confirm that the fractionation was carried out correctly. When calculated values are significantly lower than the directly measured values, degradation could have taken place in the course of fractionation.

Additionally, one needs information on fractionation efficiency for comparison and optimization of different fractionation procedures. *Fractionation efficiency* may be judged as follows:

(iv) *Non-uniformity*: When calculated number and weight averages are available according to (ii), the "non-uniformity" $U$ following Eq. (2.6) is ascertained. The greater this value $U$ for the whole polymer (calculated from the MW values of all fractions) the more efficient and reliable was the fractionation.

(v) *Shifting of MW averages* [12–14]: A measure of fractionation efficiency is the difference in MW averages of a certain fraction and of the original sample.

If one takes into consideration the mass of the fraction (which influences the MW average and the sharpness of the fraction, see Sect. 3.2), it holds for measured number averages that

$$e_n'' = (w''/w)(\bar{M}_n'' - \bar{M}_n)/\bar{M}_n \tag{4.2}$$

and

$$e_n' = (w'/w)(\bar{M}_n - \bar{M}_n')/\bar{M}_n \tag{4.3}$$

where $e_n''$ and $e_n'$ are the efficiency values of gel and sol phase, respectively. Analogous equations are valid for weight averages. For the fractions having highest ($\bar{M}_n''$ or $\bar{M}_n' > \bar{M}_n$) and lowest MW ($\bar{M}_n''$ or $\bar{M}_n' < \bar{M}_n$), the highest and lowest efficiency values (plus and minus, respectively) should be expected. Fractionation efficiency is the higher the higher or lower the efficiency value for a given fraction (!) is. Without consideration of the mass of the individual fractions, efficiency is equivalently characterized by

$$f_n'' = (\bar{M}_n'' - \bar{M}_n)/\bar{M}_n \tag{4.4}$$

and

$$f_n' = (\bar{M}_n - \bar{M}_n')/\bar{M}_n \tag{4.5}$$

or analogous equations comprising weight averages.

(vi) *SEC curves*: SEC measurements of fractions provide information about MWD and, consequently, sharpness of a certain fraction. In addition, all MW averages of fractions can be calculated from the elution curve according to Eqs. (2.2) through (2.4) [11, 15, 16].

(vii) *Turbidimetric-titration curve*: Ascertainment of turbidimetric-titration curves of fractions is a quick and simple method to obtain information about fractionation efficiency. The curves in the case of a good fractionation are steep and nearly form a straight line. The onset of turbidity ($\varphi^*$) differs widely for individual fractions. Details of this method are given in Sect. 5.4.2.

### 4.3.2 Development of Molecular Weight Distributions

The development of MWDs is usually not the aim of fractionations on a preparative scale. Nevertheless, the evaluation of data according to the MWD can be important since the MWD yields the best information about the course and the efficiency of the fractionation. A defective fraction, e.g., a wrong MW determination, can be easily detected.

If at least approximately ten fractions and the corresponding MW or DP are obtained, one can proceed to evaluate MWD (cf. Sect. 2.1) [4, 17–21]. Data are summarized in a list as in Table 4.3. Here, we use DP instead of MW.

**Table 4.3.** List of fractionation data (schematically)

| Fraction number i | $w_i$ (g) | $w_{rel,i}$ [a] (%) | $\sum w_{rel,i}$ [a] (%) | $I(P_i)$ [a,b] (%) | $P_i$ |
|---|---|---|---|---|---|
| 1 | 1.4534 | 13.4 | 13.4 | 6.7 | 125 |
| 2 | 0.8733 | 8.0 | 21.4 | 17.4 | 230 |
| 3 | 1.2453 | 11.4 | 32.8 | 27.1 | 335 |
| 4 | 1.2335 | 11.3 | 44.1 | 38.5 | 450 |
| 5 | 1.2304 | 11.3 | 55.4 | 49.8 | 600 |
| 6 | 1.1789 | 10.8 | 66.2 | 60.8 | 680 |
| 7 | 1.2731 | 11.7 | 77.9 | 72.0 | 845 |
| 8 | 0.9747 | 8.9 | 86.8 | 82.3 | 1085 |
| 9 | 1.4153 | 13.2 | 100.0 | 93.4 | 1450 |
| | 10.8779 | | | | |

[a] Instead of percentage unit, one can also use mass fractions.
[b] Calculated according to Eq. (4.6).

At first, it is convenient to transform absolute masses of fractions into relative values (percentage or mass fraction; this is a reduction in information, thus, it is a good practice to report at least the absolute yield of the fractionation as well). The masses of the individual fractions, $w_i$, should be similar to reach a good gradation in MW and an optimal evaluation of data (see below). In our schematic example, 11.1% is the result of $w_{rel,i}$. This is fulfilled very well for the inner fractions 3 through 7 whereas it is sufficient for the remaining fractions. (In all honesty, one can be content with an experimental fractionation where $6\% < w_{rel,i} < 18\%$.)

Starting with fractionation data, development of MWD generally follows the scheme given in Table 4.4:

(i) *Bar graph*: The bar graph is a steplike, cumulatively plotted fractionation curve that gives an immediate picture of the course of fractionation (masses, MW, and balance of fractions). To obtain this graph, quantities $w_{rel,i}$ are cumulatively plotted vs. $P_i$, see Fig. 4.1. These bars provide only a rough approximation of the actual DPD because the distributions of the individual fractions are neglected. To overcome this, the cumulative (integral) mass distribution function $I(P)$ (cf. Sect. 2.1) is introduced by transformation of the discontinuous bar graph into a continuous distribution.

(ii) *Cumulative mass distribution function* $I(P)$: $I(P)$ delineates (in an ideal case) the mass centers of all bars of the inner fractions. These points can be calculated according to Schulz and Dinglinger [19]:

$$I(P_i) = w_{rel,i}/2 + \sum_{j=1}^{i-1} w_{rel,j} \tag{4.6}$$

**Table 4.4.** Development of molecular weight distributions

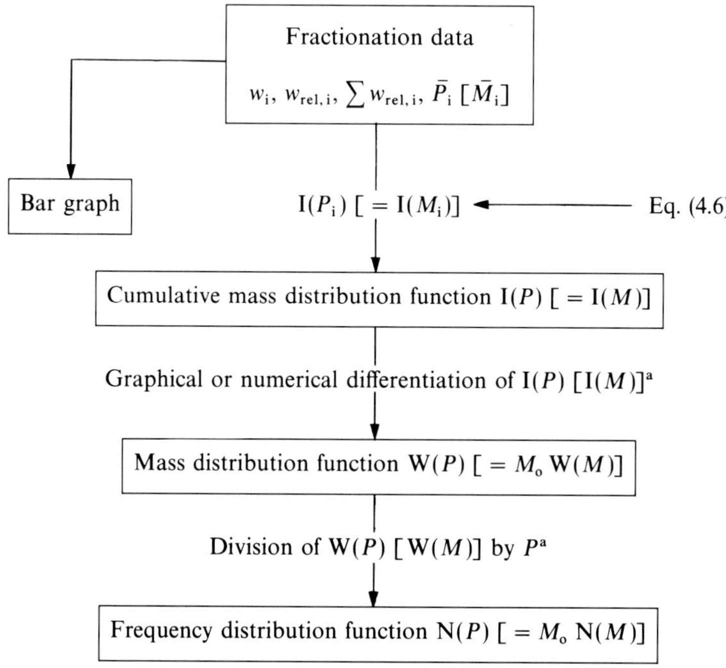


[a] Nowadays, mass distribution and frequency distribution functions are only seldom determined by fractionation. SEC is usually used.

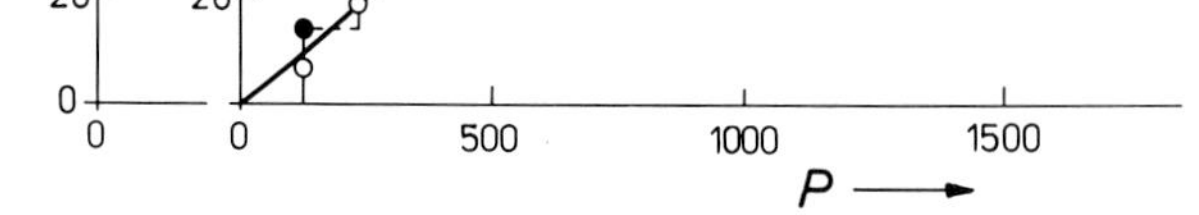


**Fig. 4.1.** Bar graph (related to $\sum w_{rel,i}$) and cumulative mass distribution function I($P$) of the fractionation summarized in Table 4.3. ● – $\sum w_{rel,i}$ for fraction 1 to i; ○ – values of I($P$) calculated according to Eq. (4.6)

The values are shown in Table 4.3. Inspection of the I($P$) curve in Fig. 4.1 shows a good agreement of the curve and the mass centers of the bars for most fractions. This results from the adjusted balance of the masses of individual fractions. Three deviations are noticeable. The fifth fraction deviates. This may be caused by two factors: (i) The determination of $P_i$ is wrong or, (ii) the

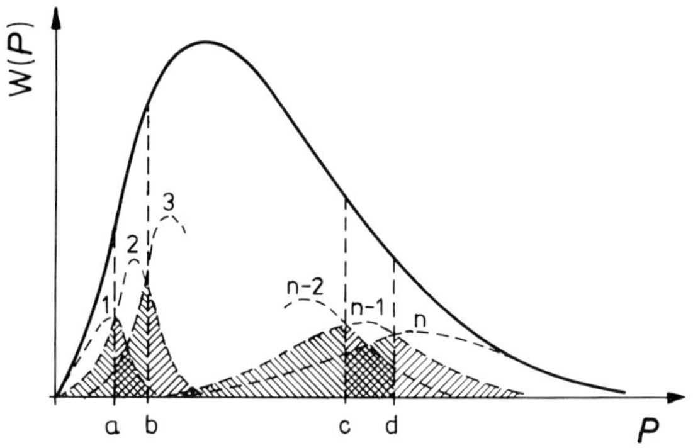


**Fig. 4.2.** Schematic graph of the effect of fraction overlapping based on calculations given by Schulz [19] (see text)

fractions are unequal in their masses. Here, the first factor holds true and we should neglect this deviation.

Two other deviations are seen for the first and last fraction. These, however, are related to the problem of *overlapping of the DPD of the individual fractions* (Fig. 4.2):

Let us assume a sharp separation of fractions symbolized by vertical lines (a – d). For actual fractions, however, overlapping occurs. In the sharp fraction 2, the hatched areas left from *a* and right from *b* are missing (because these molecules are contained in fractions 1 and 3, respectively). This loss is nearly symmetrical compensated by the hatched areas right from *a* and left from *b*. Therefore, the measured average $P_i$ represents correctly this fraction. The same holds for fraction n – 1 and the hatched areas right and left from *c* and *d*, respectively. However, the sharp fraction 1 is reduced by the hatched area right from *a* while the hatched area left from *a* occurs additionally. Thus, in comparison with the actual fraction 1, higher DP values are replaced by lower values, but, here only in the right flank of the fraction. As a result, the measured DP (of the fraction with lowest DP) shifts to a somewhat higher value than the true one for the sharp fraction. In terms of the bar graph, which represents sharp fractions, the mass center of fraction 1 should be connected with a lower $P_i$ value. The opposite effect is displayed for fraction n with highest DP. Therefore, DP shifts in the opposite direction. The curve I($P$) must compensate these deviations as Fig. 4.1 shows.

Let us return to the *mass balance of the complete fractionation* mentioned in Sect. 4.3.1. If one finds differences in the masses of the starting polymer and the sum of fractions after sufficient drying, the question is how the difference should be distributed to the individual fractions. A general answer is not possible due to many reasons. Thus, only some possibilities can be discussed:

- If it is reasonable to assume that some of the polymer has been lost in all fractions to nearly the same extent by operations such as phase isolation, redissolution, precipitation, or filtration, one should proceed with the true masses of the fractions. The same holds good if we find a too high sum of

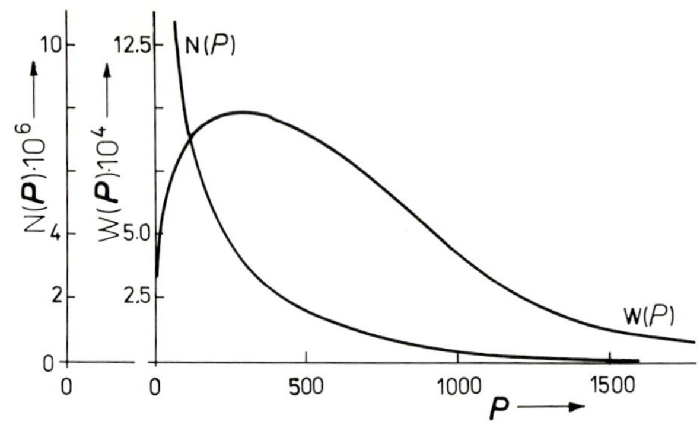


**Fig. 4.3.** Mass distribution function W($P$) and frequency distribution function N($P$) of the fractionation shown in Fig. 4.1

masses due to elution of non-polymeric material from a support in elution fractionation.

- If it is plausible that loss of the polymer is caused by incomplete reprecipitation of low-molecular fractions, the loss should be ascribed to these fractions.
- Please, keep in mind, however, that loss of polymer and insufficient drying of the fractions can compensate each other or an insufficiently dried starting polymer can simulate loss of polymer.

In any case, a fair interpretation of fractionation data requires the knowledge of the absolute, uncorrected masses of the fractions. These should be given, even if further evaluation is performed with $w_{\mathrm{rel,i}}$ data.

Finally, it should be pointed out that for equivalent values of $P$ and $M$, I($P$) is equal to I($M$) (cf. Table 2.1).

(iii) *Mass distribution function W(P)*: According to Table 2.1, one obtains the differential mass distribution function W($P$) by differentiation of the curve I($P$). This can be performed graphically point by point or numerically. A simple procedure for the latter consists in the subdivision of the curve in sufficiently small segments described by $\Delta P$ and $\Delta$I($P$) intervals. The result of differentiation of the example given in Table 4.3 and Fig. 4.1 is plotted in Fig. 4.3.

It should be noticed that functions W($P$) and W($M$) are not equal owing to the abscissa factor $M_0$ linking I($P$) and I($M$) (cf. Sect. 2.1).

(iv) *Frequency distribution function N(P)*: Frequency distribution function N($P$) can be obtained by division of W($P$) by $P$.

### 4.3.3 Evaluation of Fractionation Data by Distribution Functions

Generally it is possible to derive DPD from polymerization statistics [22, 23]. However, the "natural" polymerization statistics is increasingly influenced by technological manipulations to improve the properties of resulting polymers. Therefore, the practical importance of theoretical distribution functions decreases. Here, three distribution functions will be discussed briefly. Their formalism is given in the Appendix (Tables A 5 to A 7). Details of distribution functions can be found in the literature [17, 18, 20, 24, 25].

(i) *Generalized Schulz distribution* (Table A 5): The generalized Schulz distribution is the result of statistical derivation for polymerization and polycondensation reactions (Schulz [22], Flory [23]) and of generalization steps (Schulz [26], Zimm [27]). The cumulative Schulz function I($P$) can be calculated numerically.

(ii) *Tung distribution* [20, 28] (Table A 6): DPDs often depend not only on reaction statistics but also on thermic or other technological influences. Therefore, empirical distribution functions such as the Tung distribution are of practical importance to fit fractionation results. Unlike the Schulz distribution, W($P$) can be integrated analytically.

(iii) *Logarithmic normal distribution* (*LND*) (Table A 7): This distribution is often suitable for description of MWD with a relatively wide range of high MWs. LND was introduced in polymer chemistry by Lansing and Kraemer [29] and successfully applied by Wesslau [30] to describe the MWD of polyethylenes. If one plots LND with a logarithmic MW scale for the abscissa, a gaussian "bell-shaped" curve results. Integration of W($P$) is possible only numerically using the gaussian integral. Graphic evaluation of the integral distribution I($P$) is easy by cumulative plotting of fractionation data in a sum-probability grid (according to the gaussian integral for I($P$)) with logarithmic MW abscissa. Here LND gives a straight line.

In addition to these three types of distributions, some others are sometimes used for evaluation of fractionation data (see the literature given above).

## 4.4 Fractionation of Copolymers

Apart from MW, solubility of copolymers is also influenced by CC. However, this dependence is usually not a linear function, i.e., does not follow a straight line linking the solubilities of the parent polymers (homopolymers). Mostly, copolymers are more soluble than corresponding homopolymers. Constitution of copolymer molecules (cf. Table 1.1 and Glossary) may additionally influence the solubility. For example, solubility of random and block copolymers of the same overall CC is different in certain solvents or solvent/nonsolvent systems.

Thus, copolymer composition and constitution must affect fractionation based on solubility. One has to distinguish two cases:

(i) The effect of composition is, in a certain range of composition and with a suitable solvent/nonsolvent combination, negligible against the effect of MW ("homogeneous" copolymer). Then, fractionation can be carried out analogously to the fractionation of a homopolymer since the MW dependence of the solubility is only shifted according to CC.

(ii) If solubility and fractionation are influenced by both MWD and CCD ("heterogeneous" copolymer), temperature variation or use of different solvent/nonsolvent systems in fractionation procedures lead to different results. Therefore, one usually obtains fractions different in MW *and* CC. Both apparent MWD and CCD result in being too narrow.

The problem of copolymer fractionation was quantitatively described by Topchiev et al. [31] starting from Flory-Huggins theory (cf. p. 12). It follows a relation for the distribution coefficient $K_i$ of copolymer species with $P_i$ and composition $w_i$ in analogy to Eq. (3.8):

$$K_i = \exp[P_i(\sigma + \kappa \cdot w_i)] \tag{4.7}$$

$\sigma$ and $\kappa$ are fractionation parameters depending on the conditions of fractionation. $\sigma$ summarizes volume fractions and Huggins coefficients of the components (copolymer, solvent, nonsolvent) in sol and gel phases. $\kappa$ was specified by Teramachi and Nagasawa [32] for a copolymer AB in solvent(1)/nonsolvent(2) systems

$$\kappa = (\varphi_1' - \varphi_1'')(\chi_{1A} - \chi_{1B}) + (\varphi_2' - \varphi_2'')(\chi_{2A} - \chi_{2B}) \tag{4.8}$$

Equation (4.7) shows that $\kappa = 0$ enables the fractionation of copolymers solely with respect to DP (MW) whereas the influence of DP is always present in fractionation according to CC. $\kappa = 0$ holds good if $\chi_{1A} = \chi_{1B}$ and $\chi_{2A} = \chi_{2B}$. Homogeneous copolymers can be fractionated according to DP since $w_i$ is constant. Thus, $\kappa$ must not be zero but constant when $K_i$ is only a function of $P_i$.

Fractionation with a constant average CC but graded in MW is a difficult task because testing of different fractionation systems and procedures is necessary [17, 33, 34]. Such a fractionation is possible, e.g. with S/AN copolymers having a mole-fraction of AN of about 0.32–0.44 using dichloromethane and methanol [35, 36]. The success of such an operation can be seen from the broadness of the MWD after fractionation. Generally, a MWD is more likely the higher its non-uniformity is. Different polymer/solvent/nonsolvent systems often act in different ways. Fractionation takes place preferentially either with respect to MW or to CC. To select suitable systems, *simplified rules* may help:

(i) Similar solubilities of the parent polymers in the solvent combined with a nonsolvent of similar precipitation strength for both homopolymers should mainly bring about fractionation with respect to MW.

(ii) If solubilities of the parent polymers and the precipitation effect are different, then fractionation according to CC is to be expected.

(iii) Points (i) and (ii) may be overshadowed by the effects of polarity of polymer, solvent, and (especially) nonsolvent. Therefore, when the copolymer contains polar and nonpolar monomer units, fractionation behavior can be very complex.

A simple tool for studying fractionation behavior is the *turbidimetric-titration method* (see Sect. 5.4.2). Precipitation points $\varphi^*$ (represented by the volume fraction of the nonsolvent) are determined for copolymers with different CC as well as MW in solvent/nonsolvent systems and at different temperatures, if necessary. When a series of copolymer samples with the same monomer pair is available, the dependence of the precipitation point on MW and on CC can be determined. A measure of fractionation according to MW is the slope B in the empirical relation [37]

$$\varphi^* = \mathrm{A} + \mathrm{B} \cdot M^{-1/2} \tag{4.9}$$

Equation (4.9) holds good for most polymer/solvent/nonsolvent systems. For copolymers, this relation refers to constant polymer composition. The dependence of $\varphi^*$ on CC is very often nonlinear. A plot of $\varphi^*$ vs CC for samples with similar MW is required. It is also possible to compare differences, $\Delta\varphi^*$, of the highest and lowest value of $\varphi^*$ for a certain composition range. The higher the values of B and $\Delta\varphi^*$, the more the fractionation system works with respect to MW and CC, respectively.

Table 4.5 and Fig. 4.4 give an example for a study with poly($\alpha$-methyl styrene-*co*-acrylonitrile) in several solvent/nonsolvent mixtures [38]. The highest value of B was found with the system MEK/methanol, the lowest value with THF/*n*-hexane. The first system should be predestinated for fractionation according to MW, the last one with respect to CC. Let us first inspect the $\Delta\varphi^*$ values and Fig. 4.4. The highest values were determined in systems with *n*-hexane. According to this result and the relatively small B values, these systems could be used for fractionation with respect to CC. Please, note that the influence of MW upon fractionation is not completely eliminated because B distinctly exceeds zero. For systems with methanol, various $\Delta\varphi^*$ values are given in Table 4.5 due to the nonlinear dependencies (cf. Fig. 4.4). All values are smaller than in systems with *n*-hexane but here, the actual composition range is of importance: Fractionation according to MW of a sample with AN content (mole-fraction, $x_{\mathrm{AN}}$) between 0.30 and 0.45 should be performed with THF/methanol despite the low B value since the influence of composition in this case is negligible (more than in MEK/methanol). For a sample with a higher AN content, MEK/methanol should be chosen.

**Table 4.5.** Dependence of solubility on molecular weight and chemical composition of poly(α-methyl styrene-*co*-acrylonitrile) in different solvent/nonsolvent systems (results from turbidimetric titrations [38])

| Solvent/nonsolvent | B[a] | $\Delta\varphi^{*b}$ for $x_{AN}$ (mole-fraction) | | |
|---|---|---|---|---|
| | | 0.30–0.45 | 0.30–0.55 | 0.45–0.55 |
| MEK/methanol | 31.4 | 0.055 | 0.070 | 0.015 |
| THF/methanol | 19.0 | 0.010 | 0.050 | 0.040 |
| MEK/*n*-hexane | 20.0 | 0.155 | 0.155 | 0.155 |
| THF/*n*-hexane | 16.6 | 0.180 | 0.180 | 0.180 |

[a] B is the slope in Eq. (4.9).
[b] $\Delta\varphi^*$ represents the difference in highest and lowest precipitation point (volume fraction of nonsolvent) for a given difference in CC of copolymers having the same MW (cf. Fig. 4.4).

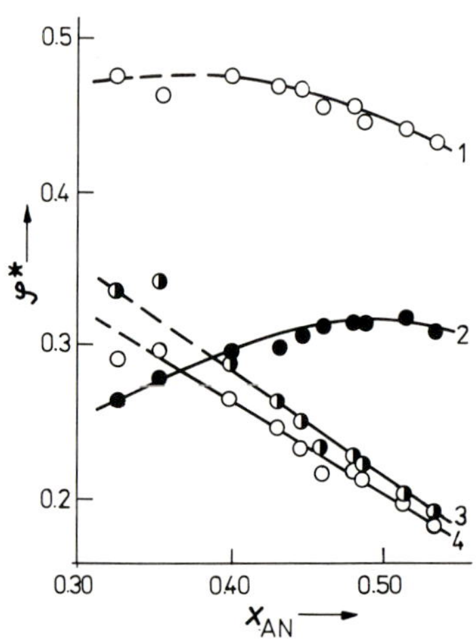


**Fig. 4.4.** Dependence of precipitation points $\varphi^*$ (volume fraction of nonsolvent) on chemical composition of αMS/AN copolymers with $\bar{M}_w \approx 10^5$ g/mol determined by turbidimetric titration. Solvent/nonsolvent systems: *1* – THF/methanol, *2* – MEK/methanol, *3* – THF/*n*-hexane, *4* – MEK/*n*-hexane. (Adopted from Ref. [38] with permission of Deutscher Verlag für Grundstoffindustrie GmbH, Leipzig)

A quantitative description of fractionation rules for copolymers starting from turbidimetric titrations is given by Stejskal and Kratochvíl [39].

A second, more precise possibility of investigating fractionation systems for copolymers consists of the fractionation of a copolymer mixture. This is illustrated by the following example [35]:

Poly(styrene-*co*-AN) 1: $x_{AN} = 0.246$, $\bar{M}_n = 62000$ g/mol

2: $x_{AN} = 0.376$, $\bar{M}_n = 44000$ g/mol

Gradient-elution fractionation (see Sect. 7):
System 1: dichloromethane/methanol
System 2: dichloromethane/*n*-hexane

Result:
System 1 elutes copolymer 2 before sample 1 although copolymer 1 is more soluble than copolymer 2 in pure dichloromethane. System 2 yields copolymer 1 before sample 2 although copolymer 1 possesses the higher MW.

Conclusion:
System 1 gives elution fractionation according to MW, system 2 mainly fractionates with respect to CC.

If one wishes to obtain the complex, two-dimensional distribution of a copolymer (MWD and CCD), a *cross-fractionation* is necessary (see Sect. 12). This method should also be used to obtain the CCD or to produce fractions with narrow distributions of both MW and CC.

## References


1. Hoffmann M, Schneider P (1963) In: Müller E (ed) Houben-Weyl Methoden der Organischen Chemie, vol 14/2. Georg Thieme, Stuttgart, p 917
2. Haslam J, Willis HA (1965) Identification and analysis of plastics. Iliffe, London
3. Schröder E, Franz J, Hagen E (1976) Ausgewählte Methoden zur Plastanalytik. Akademie-Verlag, Berlin
4. Hoffmann M, Krömer H, Kuhn R (1977) Polymeranalytik I. Georg Thieme, Stuttgart
5. Hummel DO, Scholl F (1978, 1981, 1984, 1988) Atlas of polymer and plastics analysis, 2nd edn. Carl Hanser, München; VCH Verlagsgesellschaft and VCH International, Weinheim, vol 1 (1978), 2a (1984), 2b (1988), 3 (1981)
6. Fuchs O (1989) In: Brandrup J, Immergut EH (eds) Polymer handbook, 3rd edn. Wiley, New York, p VII/379
7. Crompton TR (1989, 1991) Analysis of polymers. Pergamon, Oxford
8. Kamide K (1989) In: Booth C, Price C (eds) Comprehensive polymer science, vol 1. Pergamon, Oxford, p 75
9. Katime IA, Quintana JR (1989) In: Booth C, Price C (eds) Comprehensive polymer science, vol 1. Pergamon, Oxford, p 103
10. Lovell PA (1989) In: Booth C, Price C (eds) Comprehensive polymer science, vol 1. Pergamon, Oxford, p 173
11. Dawkins JV (1989) In: Booth C, Price C (eds) Comprehensive polymer science, vol 1. Pergamon, Oxford, p 231
12. Huggins ML, Okamoto H (1967) In: Cantow MJR (ed) Polymer fractionation. Academic Press, New York, p 1
13. Koningsveld R (1970) Adv Polym Sci 7: 1
14. Koningsveld R, Kleintjens LA, Geerissen H, Schützeichel P, Wolf BA (1989) In: Booth C, Price C (eds) Comprehensive polymer science, vol 1. Pergamon, Oxford, p 293
15. Tung LH, Moore JC (1977) In: Tung LH (ed) Fractionation of synthetic polymers. Marcel Dekker, New York, p 545
16. Glöckner G (1987) Polymer characterization by liquid chromatography. Elsevier, Amsterdam
17. Schröder E, Müller G, Arndt KF (1988) Polymer characterization. Hanser Publ, Munich; (1989) Polymer characterization. Akademie-Verlag, Berlin
18. Peebles Jr LH (1971) Molecular weight distribution in polymers. Polymer Reviews (eds: Franke HF, Immergut EH), vol 18. Wiley-Interscience, New York

19. Schulz GV (1953) In: Stuart HA (ed) Die Physik der Hochpolymeren, vol II. Springer, Berlin Göttingen Heidelberg, p 726 (Chapter 17)
20. Tung LH (1967) In: Cantow MJR (ed) Polymer fractionation. Academic Press, New York, p 379
21. Goodrich FC (1967) In: Cantow MJR (ed) Polymer fractionation. Academic Press, New York, p 415
22. Schulz GV (1935) Z Phys Chem B 30: 379
23. Flory PJ (1936) J Amer Chem Soc 58: 1877
24. Booth C, Colclough RO (1989) In: Booth C, Price C (eds) Comprehensive polymer science, vol 1. Pergamon, Oxford, p 55
25. Tang AQ (1985) Statistical theory of polymer reactions. Academic Press, Beijing
26. Schulz GV (1939) Z Phys Chem B 43: 25
27. Zimm BH (1948) J Chem Phys 16: 1099
28. Tung LH (1956) J Polym Sci 20: 495
29. Lansing WD, Kraemer EO (1935) J Amer Chem Soc 57: 1369
30. Wesslau H (1956) Makromol Chem 20: 111
31. Topchiev AV, Litmanovich AD, Shtern VYa (1962) Dokl Akad Nauk SSSR 147: 1389
32. Teramachi S, Nagasawa S (1968) J Macromol Sci, Chem A2: 1169
33. Fuchs O, Schmieder W (1967) In: Cantow MJR (ed) Polymer fractionation. Academic Press, New York, p 341
34. Riess G, Callot P (1977) In: Tung LH (ed) Fractionation of synthetic polymers. Marcel Dekker, New York, p 445
35. Glöckner G, Francuskiewicz F, Müller KD (1971) Plaste & Kautschuk 18: 654
36. Glöckner G, Francuskiewicz F, Müller S (1975) Faserforsch Textiltechn 26: 287
37. Glöckner G (1965) Z Phys Chem 229: 98
38. Glöckner G, Francuskiewicz F, Reichardt HU (1979) Plaste & Kautschuk 26: 431
39. Stejskal J, Kratochvíl P (1978) Macromolecules 11: 1097

# 5 Precipitation Fractionation

Let us now consider practical aspects of different fractionation procedures based on change of solubility:

- Precipitation fractionation (this section)
- Extraction fractionation (without continuous gradient, Sect. 6)
- Gradient-elution fractionations with solvent or temperature gradient (Sects. 7–10)
- Fractionation by partition between immiscible or demixing solvents (Sect. 11).

A survey of fractionations described in the literature for various polymers and techniques is given in Refs. [1, 2].

## 5.1 Principles and Limitations of Application

The principle of precipitation fractionation depends on phase separation governed by a decrease in the dissolution power of the solvent used [3, 4]. This can be implemented in several ways:

- Addition of a nonsolvent (precipitant) increases the $\chi$ parameter (cf. Sect. 3.1) and leads to phase separation into the sol and gel phases.
- The same effect is accomplished by preferential evaporation of the solvent in a solvent/nonsolvent mixture.
- The Huggins constant $\chi$ usually increases with decreasing temperature leading to phase separation. In systems having LCST behaviour (see Glossary), an increase of temperature gives the same result. Pure liquids and mixtures of solvent and nonsolvent can be used as solvents.

In all cases, one obtains a sol and a gel phase which can be used to obtain the desired fraction. In normal precipitation fractionation ("downward", see p. 16), gel phases (containing about 10 percent of the starting polymer) are isolated and processed further, whereas sol phases are used for the next fractionation step. In the "upward precipitation fractionation" (see p. 16), the opposite holds: Sol phases (now containing about 10 percent of the starting polymer) deliver the desired fractions, and the gel phases return to the fractionation process (see Table 5.2). Consequently, one obtains the first fraction with the lowest MWs, whereas in normal precipitation fractionation, the first fraction contains the polymer with the highest MWs.

After the theoretical considerations given in Sects. 3 and 4, some important aspects of the practical realization of precipitation fractionations will be repeated:

- The dissolution strength of the solvent and the precipitation strength of the nonsolvent must be not too high (cf. Sect. 4.1.2).
- The concentration of the starting polymer solution influences the fractionation efficiency strongly (cf. Sect. 3.2) and should be relatively low.
- Rapid and equal adjustment of precipitation equilibrium is necessary.
- The smaller the mass of the fraction in relation to the starting polymer the higher the sharpness of this fraction (cf. Sect. 3.2). Therefore, one should try to produce fractions with similar masses to obtain equally narrow fractions.
- Fractionation should be performed at temperature as near as possible to room temperature. Low temperature promotes phase separation.

A more detailed description of these points will be given in Sect. 5.3. Precipitation fractionation possesses two general features in comparison to other techniques discussed in this book: It can be carried out easily without a lot of preceding tests. However, it is a very time-consuming procedure due to the phase-separation steps. The variants of precipitation mentioned above are stamped with advantages and disadvantages.

**"Downward" Fractionation by Addition of Precipitants**

**Advantages**
- Very simple equipment and easy handling (see Sects. 5.2 and 5.3)
- Precipitation, phase separation and isolation at constant temperature
- Good control of precipitated mass of polymer via volume of precipitant when precipitation behaviour of the system is known.

**Disadvantages**
- Dilution of the system during proceeding fractionation due to addition of nonsolvent. This results in a disadvantageous course of polymer concentration (contrary to fractionation theory since highest MWs are precipitated from highest concentrations and vice versa).
- It is difficult to obtain narrow fractions of high MW.
- Increase in the fractionation volume with each fractionation step. One starts with a relatively small volume of solution that contains only a small quantity of polymer.
- A gradient of precipitant concentration occurs near the precipitant inlet.

Despite these disadvantages, "downward" precipitation fractionation is the most used precipitation procedure. It may be considered as starting point for all other fractionation procedures.

**"Upward Precipitation Fractionation"**

**Advantages**
- Use of higher polymer concentrations (cf. Sect. 3.2)
- General possibility of obtaining sharp fractions with high masses of polymer via sol phases
- Adjustment of the optimal concentration for the appropriate MW by redissolution of gel phases.

**Disadvantages**
- Additional expense due to the processing of sol *plus* gel phase in each fractionation step
- Difficulties in controlling the mass of fractions, especially in the beginning, owing to the large amount of the gel phase.

Successful "upward precipitation fractionation" requires experience and knowledge about the course of precipitation in the system studied.

**Solvent Evaporation Method**

**Advantages**
- Volume decreases with advancing fractionation.
- Better utilization of the volume of the fractionation flask makes it possible to use large samples of starting polymer for fractionation.
- Nearly constant (or only slightly decreasing) polymer concentration during the course of fractionation (i.e, with decreasing MW).
- Equal adjustment of precipitation equilibrium.

**Disadvantages**
- Only possible in systems where the boiling point of the solvent is lower than that of the nonsolvent
- Temperature control is more complicated at higher temperatures or when using vacuum.

Up till now, this method has seldom been used.

**Fractionation by Cooling**

**Advantages**
- Constant solvent composition
- Almost constant volume
- Good control of precipitation.

**Disadvantages**
- Fractionation temperatures considerably above or (for low-MW fractions) below room temperature
- Handling of gel fractions (isolation!) at constant temperatures other than ambient temperature
- Difficulties in achieving complete precipitation of low-MW components
- Elevated temperatures for relatively long periods which can be dangerous for the stability of polymer and solvent.

This method is also not used very often. However, it is advantageous for fractionation of polymers which are soluble only at elevated temperatures.

Sometimes, it is favourable to combine (dependent on the actual course of fractionation, MW range, or polymer concentration) steps of nonsolvent addition, solvent evaporation, and cooling. Thus, in precipitation fractionation, a great variety of procedures is possible. Details of implementation will be discussed in Sect. 5.3.

Generally speaking, it is difficult to obtain narrow high-molecular fractions. The course of fractionation can be very complicated for polymers which tend to associate or to re-crystallize in the liquid phase. Precipitation fractionation is very time consuming. This fact limits the number of fractions. Usually, one expects about ten fractions as a result of precipitation fractionations.

## 5.2 Equipment and Materials

*Equipment* for precipitation fractionation is simple. Normal standard laboratory glassware can be used for most steps. Table 5.1 gives a survey of standard

**Table 5.1.** Standard equipment for precipitation fractionation

| Step of fractionation procedure | Devices |
|---|---|
| Dissolution | – three-neck vessel (fractionation flask, dimensioned for mass of polymer fractionated) or other separate flask<br>– stirrer or shaker |
| Thermostatting | – bath filled with a medium of fractionation temperature<br>– stirrer or roll-round pump<br>– temperature control unit |
| Precipitation | – fractionation flask (see Fig. 5.1 and text)<br>– tight-fitting stirrer<br>– reflux condenser with drying tube<br>– dropping funnel (only for nonsolvent addition)<br>– equipment for distillation (only for solvent evaporation method – see Sect. 5.3) |
| Phase isolation | – pipettes or special devices (see Sect. 5.3 and Fig. 5.2)<br>– flask to take up gel or sol phase |
| Re-dissolution | – vessel with ground-glass joint |
| Re-precipitation | – glass beaker (> 1 l)<br>– stirrer<br>– dropping funnel |
| Drying | – device for filtration (by vacuum or pressure)<br>– vacuum oven |

equipment for the principal steps in precipitation fractionation. Remarks are necessary for some steps:

- Dissolution of polymer should be performed outside the fractionation flask to separate small insoluble parts before fractionation starts. Ensure quantitative transfer!
- The bath used for thermostatting should be adequate to ensure temperature constancy and careful heating and cooling.
- Several fractionation flasks are depicted in Fig. 5.1. Flasks (a), (c), and (d) make the isolation of sol and gel phases easy because of the small interface between the phases. If the gel phase tends to stick to the side of fractionation vessel, it is important that the flask possesses steep sides (flask (b)). Flask (d) makes the isolation of gel phase very elegant, but the stop-cock must be tight. In addition, some problems of phase isolation (see Sect. 5.3) may arise.
- Hints regarding special devices for phase isolation will be given in Sect. 5.3 (cf. also Fig. 5.2).
- Filtration of re-precipitated polymer mostly calls for a vacuum or pressure. Sometimes, precipitated polymer particles can be so small that centrifugation must be used instead of filtration.

Let us now discuss the *amounts of polymer, solvent, and nonsolvent* necessary in precipitation fractionation.

**Mass of Starting Polymer**

- Normal fractionation should start with polymer concentrations of 0.5–1.0%, i.e., one is able to fractionate 10–20 g in a 4 l flask by use of 1–2 l solvent depending on the amount of nonsolvent necessary in the course of fractionation.
- "Upward precipitation fractionation" can start with higher polymer concentrations. For example, one can dissolve about 50 g polymer in the same vessel with the same volume of solvent as in normal fractionation.
- In fractionations with a nearly constant or decreasing fractionation volume (cooling or solvent evaporation), one can use up to 30 g polymer dissolved in 3 l solvent if only a small amount of nonsolvent is necessary (4 l flask).

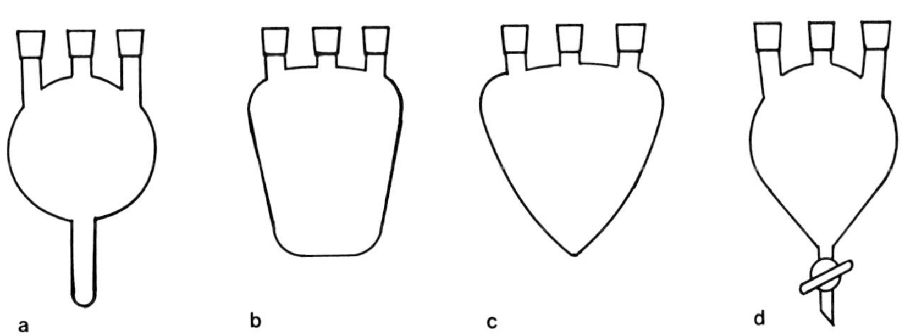


**Fig. 5.1a–d.** Different fractionation flasks for precipitation fractionation (see text)

**Volumes of Solvent and Nonsolvent**

- Except for "upward precipitation fractionation", solvent is only necessary for dissolution of the starting polymer (see above) and for re-dissolution of (relatively small) gel phases. Procedures without stepwise addition of nonsolvent need less precipitant than in the case of nonsolvent addition. In the latter case, one can assume a volume of precipitant similar or somewhat greater than the solvent volume used. Here, the precipitant necessary for re-precipitation of fractions is not included. Re-precipitation requires about a tenfold nonsolvent volume compared with the solvent used for re-dissolution of gel phases (cf. Sect. 4.2.2).
- In "upward precipitation fractionation", one needs greater volumes of both solvent and nonsolvent. In the example mentioned above, 4–5 l are required due to the alternation of precipitation and re-dissolution of all gel phases. However, if we take into consideration the mass of polymer fractionated, one needs comparable volumes related to a certain mass of polymer. The volumes for processing of the fractions after sol-phase isolation are not included.

**Guidelines**

To sum up, the following guidelines about required materials can be given:

- Solvent for dissolution of the polymer (about 100–200 ml per gram polymer)
- Solvent for re-dissolution before re-precipitation (about 100 ml per gram polymer)
- Nonsolvent for precipitation (in the sum for all fractions approximately equal to the solvent volume necessary for dissolution)
- Nonsolvent for re-precipitation (about 1 l per gram polymer), and, if necessary
- Small amounts of stabilizers.

The specifications of solvent and nonsolvent volumes given in parentheses are influenced by the strength of both solvent and nonsolvent.

## 5.3 Specific Steps of Implementation

### 5.3.1 Course of Precipitation Fractionations

Before one starts with precipitation fractionation, preliminary investigations of polymer, solvent, and nonsolvent must be carried out according to Sect. 4.1 and Tables 4.1 and 4.2. General rules given in Sect. 4.2 should be taken into consideration. A systematic survey of the course of precipitation fractionations including the variants discussed above is given in Table 5.2.

Let us now discuss the fractionation steps according to their experimental succession. Special hints and experimental tricks are included.

**Table 5.2.** Course and variants of precipitation fractionations (For preliminary investigations see Tables 4.1 and 4.2)

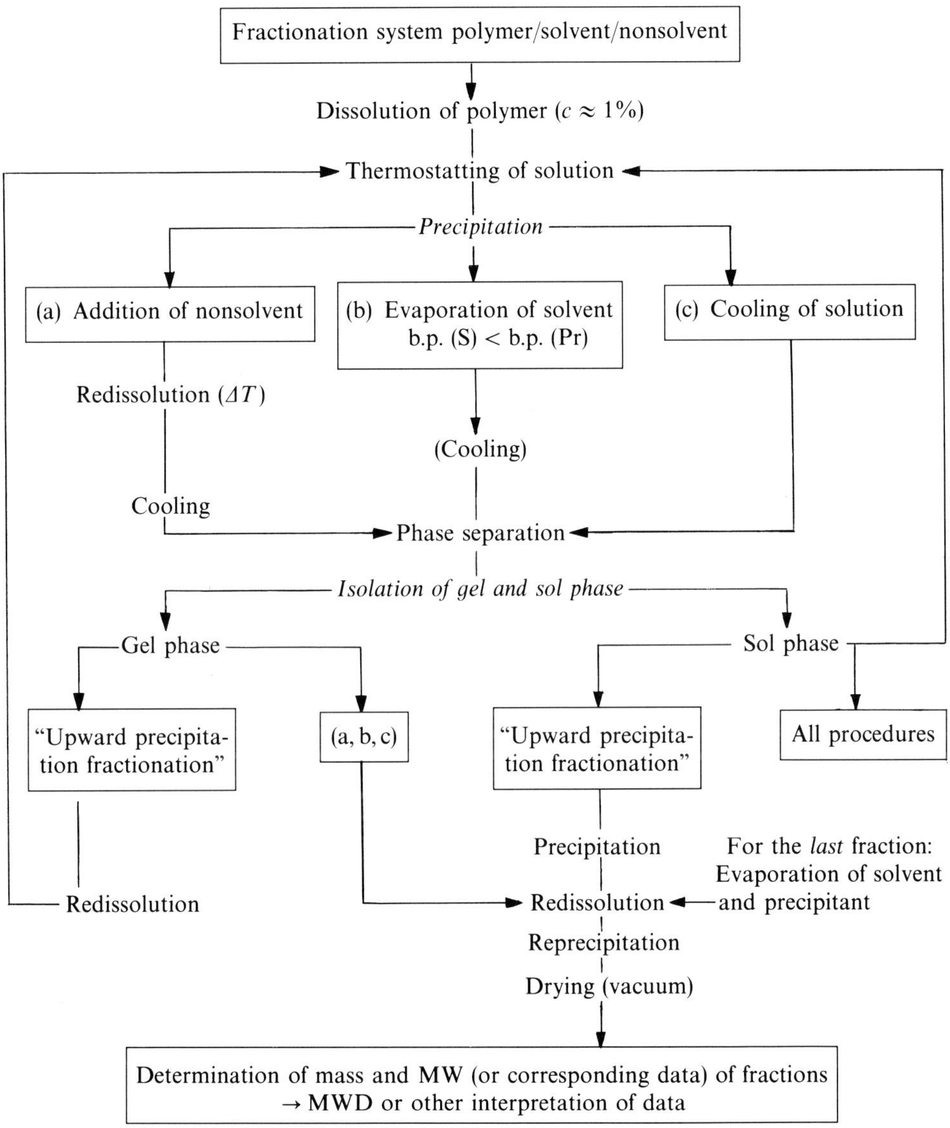


**Experimental Setup**

The fractionation vessel according to Figure 5.1a through d is equipped with a tight-fitting stirrer, a reflux condenser with drying tube, and a dropping funnel. The volume of the liquid fractionation system should not exceed about 50% of the flask volume to ensure adequate stirring. If it is necessary to seal up the ground-glass joints, gaskets of poly(ethylene) or teflon foils can be used.

*Grease must not be used* if it is soluble in solvents or precipitants. For the arrangement of stirrer and dropping funnel, one has to ensure that the precipitant drops into the solution near to the stirrer to facilitate rapid intermixing of precipitant and solution. Finally, one has to position the fractionation flask inside the thermostat so that the sol and gel phases can be isolated after separation without a change of the arrangement (see below).

**Dissolution of the Polymer** Dissolution of the polymer should be carried out outside the fractionation flask to allow for cleaning (filtration, centrifugation) of the solution. The period of dissolution should not be too short (about 12 to 24 hours) to obtain molecularly dispersed solutions. Solution concentration may range from about 0.1 to 1.5%. The concentration should be the higher the lower the MW of the solute is. Usually, a concentration of about one percent is appropriate. Lower concentrations improve the sharpness of the fractions but, simultaneously, the separation (sedimentation) of the gel phase is hampered. On the other hand, enhanced concentration can lead to a better phase separation but also to changes in the sequence of fractions ("reverse-order fractionation", see p. 17). If the polymer has to be dissolved at an elevated temperature (crystallinity or fractionation by cooling), the necessity of stabilization should be checked.

**Thermostatting** Before precipitant can be added, the solution has to be under isothermal conditions which should not change when the precipitant is added. Furthermore, it is necessary that the level of liquid in the thermostat is higher than that in the fractionation vessel in all stages of fractionation. Temperature fluctuations must be prevented by thermal insulation of the thermostat. However, the insulation must allow the observation of the precipitation process; a window should be always free. Temperature changes must not exceed $\pm$ 0.05 K. Therefore, two independent control units are recommended due to the extended time of the fractionation procedure. As a rule, the *fractionation temperature should be as near as possible to room temperature.* The higher the difference between the fractionation and room temperature, the more difficult it is to handle the sol and gel phases in the course of fractionation (see below). If one wishes to perform fractionations by solvent evaporation or cooling, variable temperature regimes and good thermal insulation are essential.

**Precipitation** **Addition of precipitant** has to proceed with permanent agitation of the solution to prevent a precipitation gradient in the polymer solution. The stirring rate should guarantee complete mixing in the system. *The precipitant must be added slowly* (1–2 droplets per second). If a meager permanent turbidity occurs, the addition of precipitant should be broken off. Continuous agitation promotes the establishment of precipitation equilibrium.

Estimation of the *amount of the gel phase* by the turbidity of the solution is not easy, because turbidity does not depend on the mass of polymer in the gel phase alone but also on the difference between the refractive indices of the sol and gel phases, on the size and shape of the droplets of the gel phase and on the thickness of the fractionation volume. It may be helpful to place a printed paper (e.g., a newspaper) behind the fractionation flask for the estimation of turbidity. Furthermore, information about the refractive indices of solvent, precipitant, and polymer is profitable.

Carrying out the precipitation process successfully requires a certain degree of experience and is not free of arbitrariness. To overcome this problem to some extent, it is possible to measure the turbidity (or extinction) of a small volume of the system. However, one has to recognize that this method and, likewise, measurements of the volume of the settled gel phase are not reliable means of obtaining fractions of equal amounts of polymer. In addition, one should try to precipitate the marginal fractions (i.e., the first and the last one) with a somewhat smaller mass fraction than the other fractions owing to the asymmetrical overlapping and the fact that the marginal fractions usually possess a broader distribution (cf. Fig. 4.2).

In the case of "upward precipitation fractionation", a large amount of polymer is precipitated in each fractionation step. Therefore, the turbidity of the system is very high and the proceeding precipitation is almost impossible to control. Consequently, it is very difficult to obtain fairly balanced sol phases to isolate desired fractions. Thus, "upward precipitation fractionation" can be recommended only after an intensive study of the precipitation behavior (determination of the mass fractions of the polymer in sol and gel phases depending on polymer concentration and nonsolvent addition).

**Precipitation by solvent evaporation** must start with addition of such an amount of precipitant that the precipitation threshold of the dissolved polymer is nearly reached. This point can be found reliably by slow nonsolvent addition up to a first meager turbidity at a temperature 1–2 K below the fractionation temperature. The solution should be just clear at fractionation temperature.

Solvent evaporation can be achieved in different ways:

- The solvent/nonsolvent mixture with the polymer is heated up to the boiling point of the solvent, and after evaporation of the intended solvent volume, the system is cooled to fractionation temperature. Both evaporation and cooling should be performed while stirring. Two disadvantages should be stated here: One has to work at elevated temperatures, and, the precipitation is not observable during evaporation but only during cooling. Thus, the size of a certain fraction can be controlled only by the solvent volume distilled.

- To overcome these disadvantages, one can use a vacuum to lower the boiling point of the solvent to the fractionation temperature. Instead of stirring, one should work with a capillary tube to bubble an inert gas through the mixture. The precipitation can be observed and can be stopped immediately by decreasing the vacuum. Additionally, one is able to control the precipitation by condensation of the evaporated solvent. Temperature control must compensate for the energy of evaporation to keep a constant fractionation temperature.
- A third method is recommended in Ref. [2] i.e. the use of a warmed gas stream to drag out the solvent. Here, the temperature of the system should also be kept constant to use growing turbidity for precipitation control. Additionally, the mass or volume of the evaporated solvent can be measured by condensation.

**Precipitation by cooling** should be performed very slowly with stirring. Precipitation control is possible via turbidity of the system.

**Re-dissolution of the Gel Phase**

After precipitation by nonsolvent addition, re-dissolution by heating of the system and subsequent cooling to fractionation temperature is recommended to achieve a higher *uniformity of the precipitated fraction*. In this way, the low-molecular polymer included during precipitation can be separated from the gel phase. During the heating process, the system has to be stirred. The necessary temperature rise is higher the more precipitant is added after the onset of turbidity and the broader the MWD of the fractionated polymer. Heating should be continued till the solution is clear again, but, as a rule, the increase in temperature should not exceed 5 K. Otherwise, too much precipitant might have been added. Low-molecular fractions often need a wider temperature range to become transparent again than high-molecular fractions.

If the system displays LCST behaviour (see Glossary), heating will not be successful and must be omitted.

In the case of fractionation by cooling (or when the polymer is not precipitated during solvent evaporation at higher temperatures), re-dissolution by heating need not be carried out.

**Cooling and Phase Separation**

After re-dissolution, the gel phase occurs again upon cooling to fractionation temperature. This step is very important for the formation of gel particles and their settling down. It must be done by *very slow cooling* for some hours under continuous agitation. Consequently, one has to pay sufficient attention to the insulation of the thermostat. Sometimes, when one works at temperatures much higher than room temperature, use of a cooling program is recommended. After reaching the fractionation temperature, agitation of the turbid system should be continued for about half an hour to promote the flocculation of the gel particles. When phase separation proceeds, the stirrer must be removed since the sedimentation process should not be disturbed by any percussion. Besides, polymer cannot settle on the stirrer.

Separation is finished when the sol phase is clear. Sometimes, one also obtains a transparent gel phase (see below). *Phase separation can be hindered by two factors*: The gel phase (sometimes a small part only) separates very slowly. Then, the only way is to centrifuge the mixture of sol and gel phases (see next step). It is possible that complete phase separation has occurred, but, the gel phase has precipitated as a more or less turbid layer at the side of the fractionation flask. Therefore, the sol phase seems to be turbid. In such a case one should check the situation by taking a small amount of the sol phase with a syringe.

**Isolation of Sol or Gel Phase**

Careful isolation of sol and gel phases from each other is necessary to reach a good fractionation efficiency. If the phases are still partly intermingled, the volume of the slightly turbid sol phase must be centrifuged. Under these circumstances, problems can arise when fractionation and centrifugation temperature are disparate. Further precipitation of polymer is possible during centrifugation at temperatures below the fractionation temperature. This fact should be recognized in the selection of this temperature. Differences in temperature adversely affect the normal isolation of sol or gel phase.

There are two general possibilities for isolation

- Removal of the gel phase. The supernatant sol phase remains in the fractionation vessel for the precipitation of the next fraction.
- Removal of the sol phase and re-dissolution of the gel phase with a small amount of solvent.

Let us discuss these general possibilities. For an experimental procedure, there are various recommendations in the literature. We will restrict ourselves to simple setups. Some of them are shown in Fig. 5.2a–e.

**Removal of the gel phase** (Fig. 5.2a and b). The gel phase can be isolated by draining into a small vessel via a stopcock (a), or by extraction through the sol phase (b) with a syringe or with a weak vacuum (vessel for the protection of the vacuum device, similar to (e)!). The advantage of these procedures is the direct isolation of the gel phase without removal of the sol phase from the fractionation flask (the sol phase stays at constant temperature). A special problem in setup (a) could be the tightness of the stopcock and ground-glass joints with regard to the liquid of the thermostat over a long period. A general problem of setups (a) and (b) is that these procedures of isolation fail when the gel phase is not completely liquid (transparent), tends to stickiness or forms a film on the side of the fractionation vessel. It must be pointed out that *such a film can be transparent*! Removal of the gel phase may lead to complications owing to the high viscosity of the gel. This causes adherence of the gel phase to sides of the stopcock (a) or wall (b). Even formation of a flow profile is possible to such a

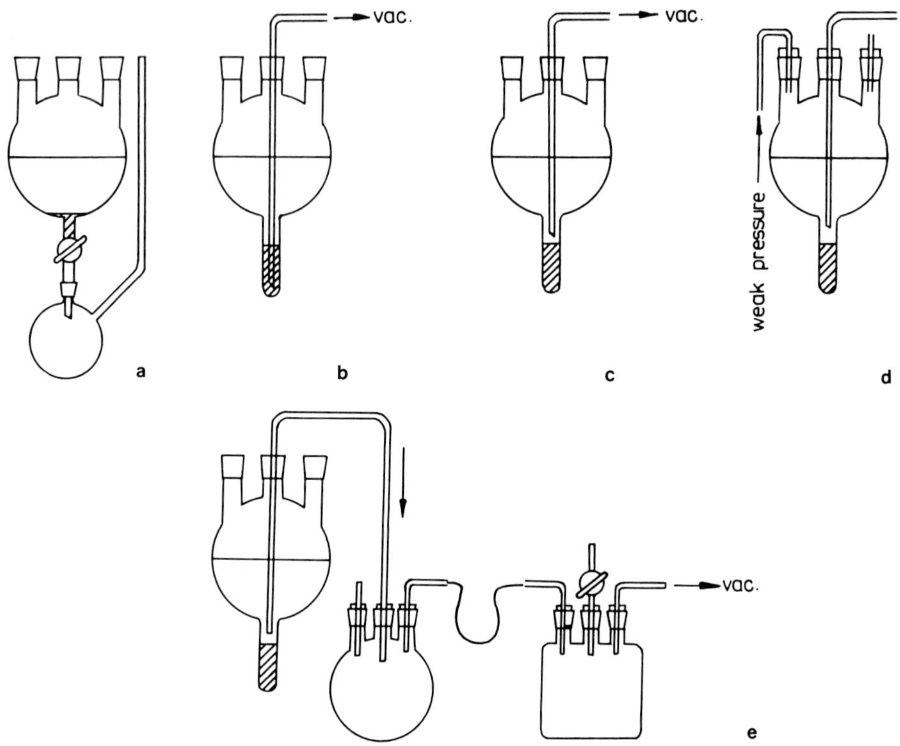


**Fig. 5.2a–e.** Possibilities for isolation of sol or gel phase. **a** and **b** – removal of the gel phase, **c**–**e** – removal of the sol phase. The vacuum equipment similar to Fig. **e** is omitted in **b** and **c** (for details see text)

degree that small amounts of sol phase go through the stopcock together with the gel phase. Hence, a complete isolation of the phases is not attainable.

**Removal of the sol phase** (Fig. 5.2c–e). The supernatant can be removed directly by vacuum (c) (equipment similar to Fig. (e) is necessary) or by special syphons (d) and (e). One is able to eliminate the problems mentioned above. However, the advantage of collecting the gel phase directly is lost. Use of (regulatable) vacuum (c) enables us, with care, to remove the sol phase completely, but there is a *possibility of evaporation of the solvent or precipitant*, at least for low-boiling liquids, owing to the vacuum.

The last-mentioned disadvantage of procedure (c) can be eliminated by use of a syphon (d) which demands a weak pressure at the surface of the sol phase. An exact separation of the phases is usually not possible with this method. Therefore, the last stage to remove the supernatant completely must be carried out often by hand with a syringe. In the setups (c) and (d), it is possible to maintain a constant temperature of the sol phase by use of a thermostat.

Procedure (e) gives the best possibility for the careful and complete separation of the phases. A moveable syphon is tightly joined to a three-neck vessel and can be filled with the sol phase by use of a vacuum when the free glass tube is closed (e.g., by a finger tip). When the syphon is working, this glass tube can be opened again, normal pressure returns, and the sol phase flows freely into the lower vessel.

**Re-dissolution and Re-precipitation Gel Phase**

Please, see Sect. 4.2.2 for these steps.

For "*upward precipitation fractionation*", the situation is somewhat different: Here, the gel phase will be re-dissolved for the sake of further fractionation. Note that the re-dissolved gel phase contains nonsolvent. The fraction of interest is obtained from the sol phase, e.g., by precipitation. For further purification, this fraction can be re-dissolved and re-precipitated once more in the same way as discussed in Sect. 4.2.2.

**Drying of Fractions**

Outlines given in Sect. 4.2.3 are valid for all fractionation procedures.

**Precipitation of Further Fractions**

Generally, all steps to precipitate any further fractions are the same as described above. Some points should be noted:

- It is possible, that the proportion of solvent and nonsolvent is changed in the remaining solution as a result of the separation procedures for sol and gel phases. A simple method of controlling the composition is the measurement of the refractive indices (assuming the difference between solvent and precipitant is high enough).
- Sometimes, particularly for precipitation of fractions with low MWs by nonsolvent addition, it might be necessary to concentrate the solution to correct for dilution in the foregoing fractionation steps. This should be done under low pressure. Again, boiling points of solvent and nonsolvent must be taken into consideration and, if necessary, solvent or nonsolvent must be added successively.
- Usually, the *last fraction* is collected by evaporation of solvent and precipitant. This fraction should be re-precipitated because of the *enrichment of pollutants or additives* in the evaporation residue. For this purpose, polymer concentration in the solution for re-precipitation should be somewhat higher than one percent.
- If re-precipitation of a fraction does not work, the polymer must be collected by evaporation. It is important to *prevent formation of a polymer film* caused by rapid evaporation of the nonsolvent. If the nonsolvent possesses the lower boiling point, it must be added successively during evaporation, particularly in the latest stage, to such an extent that a powdery precipitate is formed. This consistency permits more efficient drying of the polymer.

**Evaluation of Fractionation Data** The first results of fractionation are the masses of the individual fractions. Before one can perform further evaluation of fractionation data, measurements of MW or of related values (e.g., intrinsic viscosity) are necessary.

### 5.3.2 Reasons for Insufficient Fractionation Efficiency

**Trouble Shooting** A poor fractionation efficiency may be due to the fractionated polymer, to the fractionation conditions as well as to imperfect fractionation procedure.

**Polymer.** Sometimes, the course of fractionation is influenced not only by MWD but also by structural parameters of the individual polymer molecules like branching, ordered structures or, if one considers copolymers, by CCD. All these characteristics may influence the solubility of the polymer molecules and, therefore, the fractionation. Change of the solvent/nonsolvent system may help to minimize these problems (see Sect. 4.4).

**Fractionation conditions.** Unsuitable solvents or precipitants can be responsible for incomplete dissolution of associates or other ordered structures and may also cause coprecipitation of molecules which inherently do not belong to the very fraction. An unfavourable dissolution regime for the polymer to be fractionated (too short dissolution period, inadequate dissolution temperature) promotes the existence of associated structures of the polymer. Consequence could be inversions in the sequence of fractions, particularly for the highest-molecular fractions.

Polymer concentration influences the fractionation efficiency (see Sect. 3.2). The former strongly decreases in the course of precipitation fractionation by nonsolvent addition. As a result, the sharpness of fractions decreases with increasing MW. Therefore, for a better fractionation efficiency, one should start with relatively low concentration followed by (stepwise) consecutive solvent evaporation in the course of fractionation. Otherwise, inversions ("reverse-order fractionation") can take place.

**Imperfect procedure.** Problems may arise from incomplete phase isolation or from precipitation of too large and imbalanced fractions.

## 5.4 Modified Procedures

### 5.4.1 Sub- and Refractionation

**Subfractionation** It may happen that one or more fractions are too large after drying (say, 20 or more percent of the overall polymer mass). A *subfractionation* is recommended to obtain more balanced fractions, particularly for high-molecular fractions. The principle can be seen in Fig. 5.3a–g. Scheme (a) represents the normal precipitation fractionation, (d) illustrates the "upward precipitation fractionation". Part (b) shows schematically the course of a subfractionation related here formally for all gel fractions $G_i$. "Upward precipitation fractionation" is depicted in graph (e), where sol fractions $S_i$ can be subdivided into smaller sol and gel fractions. Of course. subfractionation is also possible by use of other fractionation procedures discussed in the subsequent sections.

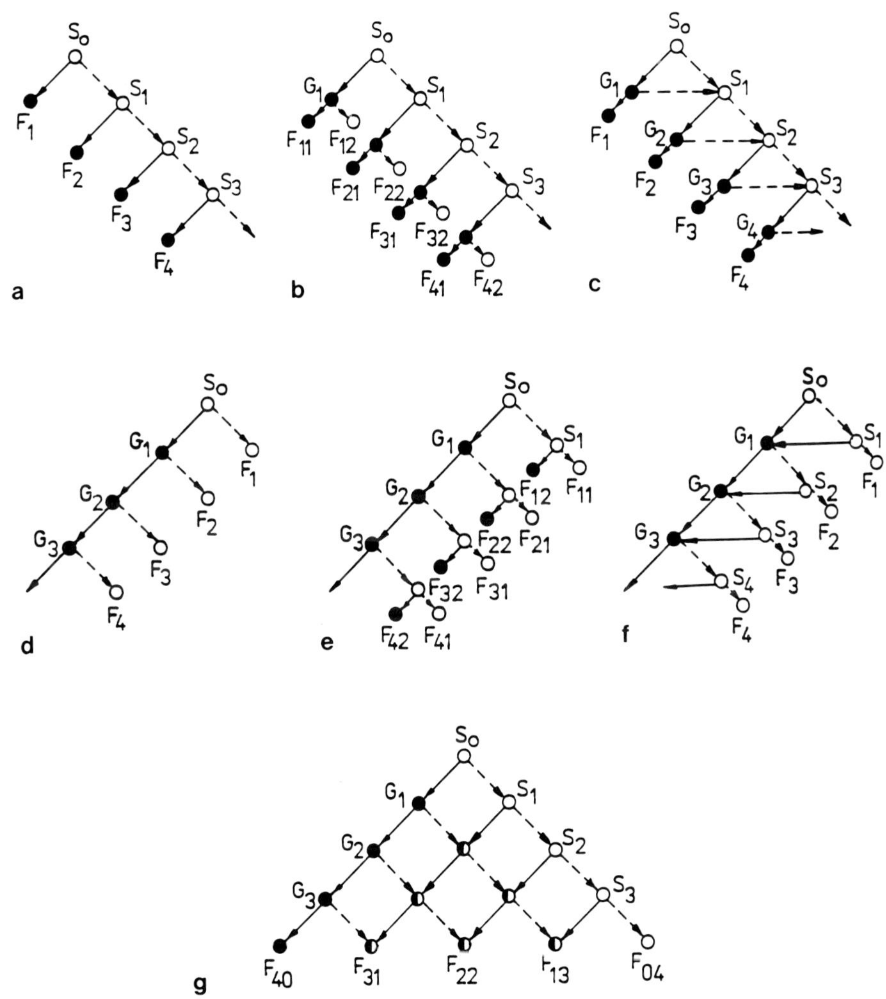


**Fig. 5.3a–g.** Fractionation schemes for "normal" precipitation fractionation **a** and "upward precipitation fractionation" **d**, extended to subfractionation **b** and **e**, and refractionation **c** and **f**, respectively. **g** represents "triangle fractionation". For details see text. →● – gel phase or fraction; →○ – sol phase or fraction; ◐ – mixed phase or fraction

**Refractionation** Another problem of fractionation is the inherent fact that each fraction possesses its own MWD and, especially in precipitation fractionation, contains an appreciable amount of lower-molecular components (low-molecular tailing, cf. Sect. 3.2). The effect occurs despite optimal precipitation equilibrium and careful phase isolation. Many proposals for *refractionation* procedures have been reported in the literature (cf. [3]) to overcome this tailing and to obtain better fractions. Schemes (c) and (f) in Fig. 5.3 show a simple variant of refractionation after Schulz [5] and Flory [6]. Other proposals are more complicated, for instance, the "triangle fractionation" of Meffroy-Biget [7] (in scheme (g) restricted to five fractions).

The schemes discussed for sub- and refractionation can also be used partly, when combined with the usual fractionation procedures. Finally, it should be emphasized that it is rather complicated to obtain reasonable concentration adjustment and mass balance for individual fractions. For example, phases $G_1$ and $S_1$ in scheme (g) should have both 50% of overall polymer mass but, $G_2$ and $S_2$ should have more than a half of $G_1$ and $S_1$, depending on the number of subsequent fractions.

### 5.4.2 Turbidimetric Methods

The turbidity of a solution can be used as a measure of the mass of polymer in the gel phase (take into account, however, remarks concerning the step "precipitation" on p. 47). Onset of turbidity (cloud point, cf. Glossary) indicates the beginning of phase separation in a solution due to nonsolvent enrichment or cooling. Various procedures generate turbidity [8–11]:

- Nonsolvent addition (continuously or discontinuously) at different rates with stirring (continuously or broken) in closed or overflow cuvettes;
- cooling at different rates with or without stirring.

Polymer concentrations should be very low (say, 20–200 mg/l) to facilitate the quantitative measurement of extinction or scattering of the turbid solution.

Cloud points are represented by the volume fraction of nonsolvent, $\varphi^*$, or by the cloud-point temperature. For a given polymer, cloud point depends upon solvent, nonsolvent, concentration, and temperature. Turbidimetric titration (i.e. addition of nonsolvent) turns out to be a powerful tool for studying solution behaviour of polymers in solvent/nonsolvent systems at low polymer concentration. The result can, after appropriate correction for dilution, be presented as curves, see Fig. 5.4. The characteristic precipitation point, $\varphi^*$, usually differs from the cloud point. Both points coincide only for sharp fractions (curve 3). Important aspects concerning turbidimetric titration are: comparison of the strength of solvents and nonsolvents, study of phase separation, influence of molecular parameters of the polymer on solubility (MW, branching, tacticity, CC).

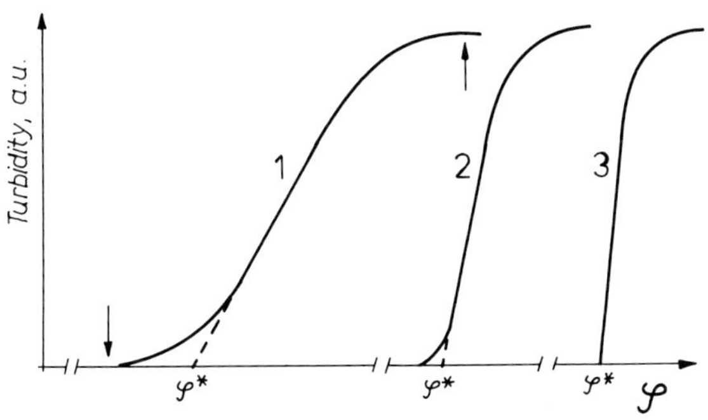


**Fig. 5.4.** Turbidimetric-titration curves of unfractionated polymer (*1*), fair fraction (*2*), and sharp fraction (*3*) (schematically). *Arrows* on curve (*1*) mark solubility limits (see text)

Turbidimetric titration is, above all, an easy method for determining binodals (see Glossary), cloud-point curves, and limits of solubility in polymer/solvent/nonsolvent systems. Cloud-point curves, binodals, and solubility limits are important for determining optimal fractionation conditions for those procedures discussed in the subsequent sections. Solubility limits are found as the first point of cloudiness (cloud point) and of total precipitation. These points are marked on curve (1) in Fig. 5.4 by arrows.

MW is connected with precipitation points according to Eq. (4.9). If one determines the constants A and B by calibration under constant conditions, MW determination is possible with only about one milligram of polymer [12]. Since $\varphi^*$ is smaller the higher the MW, advancing addition of nonsolvent produces increasing turbidity. This is caused by the increasing number of polymer molecules which cross from sol into gel phase. In other words, the course of turbidity during nonsolvent addition (turbidity curve) gives a rough picture of MWD. Turbidimetric titration is analogous to precipitation fractionation, but without separation and isolation of phases. Analogous curves result from cooling.

Turbidimetric-titration curves are different for unfractionated samples and fractions. Therefore, turbidimetric titration can be used as an easy and rapid method for obtaining information about quality (and MW, after calibration) of fractions. An example is shown in Fig. 5.4.

A special turbidimetric method for solutions having LCST behavior (see Glossary) was developed by Delmas et al. (See Ref. [13]). Turbidity is now produced by heating. This method was applied successfully for determination of MWD.

## 5.5 Examples of Precipitation Fractionations

Information given in Sects. 4.1, 4.2, 4.3.1, and 5.3.1 should be noted.

The given specifications for volumes of vessels and solvents or nonsolvents can be converted linearly for polymer masses deviating from the mentioned ones.

### 5.5.1 Precipitation Fractionation of Poly(styrene) with Toluene/Methanol by Nonsolvent Addition

**Materials**

**Polymer.** Poly(styrene) (10 g) with *rough* MW averages as follows: $\bar{M}_n = 10^5$ g/mol, $\bar{M}_w = 3\text{–}4 \times 10^5$ g/mol, $\bar{M}_v = 3 \times 10^5$ g/mol

**Solvent.** Toluene (analytical grade), 1 l for starting polymer, 100 ml/1 g polymer fraction for re-dissolution of gel phases

**Nonsolvent.** Methanol (analytical grade), 1–1.5 l for fractionation, 1 l/1 g polymer fraction for re-precipitation

**Equipment**

- Insulated thermostat (about 30 × 40 × 30 cm, filled with water) with glass window, stirrer or recirculating pump, and temperature-control unit
- Three-neck vessel as shown in Fig. 5.1a–c (volume: 4 l, for 10 g polymer)
- Tight-fitting stirrer
- Reflux condenser with drying tube
- 2 dropping funnels
- Ground-glass stoppers
- Pipettes, syringes, or devices as shown in Fig. 5.2c–e
- Flask (2 l) to take up the sol phases temporarily
- Conical flask with ground-glass joint and stopper (about 200 ml, for 1 g polymer fraction)
- Glass beaker (> 1 l, for 1 g polymer fraction to be re-precipitated)
- Stirrer
- Device for filtration by vacuum or pressure (filter of sintered glass or paper)
- Shallow glass dish
- Rotational-type evaporator (for the last fraction)
- Vacuum oven
- Stands, clamps, sockets, etc.

**Time Required**

| | |
|---|---|
| – Dissolution of starting polymer | overnight |
| – Thermostatting under agitation | 1 h |
| – Precipitation (from the first cloudiness) | 0.5–2 h |
| – Re-dissolution under agitation | 1 h |
| – Cooling under agitation | 3–4 h |
| – Phase separation | (at least) overnight |
| – Phase isolation | 1 h |

Except the dissolution of the starting polymer, all the periods are valid for each fraction. The time necessary for processing fractions is not included.

**Preparatory Investigations**

- Check of complete solubility (clear solution) of the polymer in the solvent
- Determination of the cloud point by nonsolvent addition to a small aliquote of the PS solution (10 g/l) at fractionation temperature.

**Fractionation**

Fractionation temperature: 25 °C

- Dissolution of 10 g PS with stirring or shaking in 1 l toluene. When the solution is not clear (crosslinked portions, additives, or fillers?), sedimentation, filtration, or centrifugation is necessary.
- Setting up the fractionation device: fractionation vessel with tight-fitting stirrer, reflux condenser (drying tube), and dropping funnel is put in the thermostat.
- Thermostatting of the PS solution with stirring.
- Slow addition of nonsolvent from the dropping funnel. Before the first precipitation at the cloud point (cf. preparatory investigations) is reached, nonsolvent can be added somewhat faster. The addition is broken off after the first turbidity appears (20 or 30 minutes with stirring) and is then slowly continued until strong turbidity is reached. The nonsolvent volume necessary for each fraction may change from about 8 to 10 ml for the highest-MW fraction (counted from the first cloudiness) up to more than 100 ml for the lowest-MW fractions.
- Re-dissolution of the system by heating (about 5 K) with stirring
- Slow cooling to fractionation temperature (insulation of the thermostat!) with stirring. Stirring must be continued for about 30 minutes. After this, stirrer, reflux condenser, and dropping funnel are replaced by glass stoppers.
- Separation of the phases at constant temperature. If the sol phase still seems to be turbid after a longer time, check with a small aliquote of the sol phase the reason for the apparent turbidity and possibilities for clarification.
- Isolation of the sol phase by use of syringes or devices as shown in Fig. 5.2c–e.
- Re-dissolution of the gel phase and rinsing of the fractionation vessel with a small volume of solvent; collection of the fraction in the conical flask
- Return of the sol phase to the fractionation vessel, thermostatting, and precipitation of the next fraction.
- Re-precipitation of the re-dissolved gel fraction with methanol (tenfold volume) using glass beaker, stirrer, and dropping funnel.
- Sedimentation of the re-precipitate; replacement of the toluene/methanol mixture by fresh methanol (standing overnight).
- Sedimentation and filtration (or centrifugation); drying in a glass dish under vacuum (about 60 °C, each fraction separately).
- Isolation of the last fraction by use of the rotational-type evaporator; re-dissolution, re-precipitation, filtration, drying.

The number of fractions and the added volume of methanol in each fractionation step must be estimated in connection with the masses of the foregoing (dried) fractions.

**Evaluation**

- Determination of mass and MW or related measures ($[\eta]$, $\varphi^*$, $V_e$ from SEC) for each fraction.
- Mass balance of the fractionation and of the fractions mutually. A plot of the cumulative mass of fractions vs nonsolvent volume (*precipitation curve*) allows one to optimize a subsequent repetition of the fractionation.
- Comparison of directly measured and calculated averages of MW or $[\eta]$ for the whole sample according to Eqs. (2.2) to (2.4) and (4.1).
- Comparison of gradation and sharpness of fractions and starting polymer by SEC or turbidimetric titration.

If necessary, the following evaluations may be added (cf. Sects. 4.3.2 and 4.3.3):

- Bar graph
- Distribution curves I($M$) and W($M$)
- Distribution functions.

A preparative precipitation-fractionation of high quality must be prepared by test fractionations on an analytical scale under the same conditions.

### 5.5.2 Precipitation Fractionation of Poly(styrene-*co*-acrylonitrile) with Butanone/Methanol by Nonsolvent Addition

**Materials**

**Polymer.** Poly(styrene-*co*-acrylonitrile) (10 g) with similar MW averages as mentioned for PS in Expl. 5.5.1. The copolymer should be an azeotropic one or should have a composition range (mole fraction AN) of about 0.35–0.5.

**Solvent.** Butanone (analytical grade), 1 l for starting polymer, 100 ml/1 g polymer fraction for re-dissolution of gel phases.

**Nonsolvent.** Methanol (analytical grade), 1–1.5 l for fractionation, 1 l/1 g polymer fraction for re-precipitation.

The suggested copolymer composition and the solvent/nonsolvent system conduct to fractionation according to MW. Other solvent/nonsolvent combinations or copolymers with clearly other S/AN ratio may lead to fractionation with respect to CC (see Sect. 4.4).

**Equipment** The poly(styrene-*co*-acrylonitrile) can be fractionated using the same equipment as described for the fractionation of poly(styrene) because the same procedure is used. For details see Sect. 5.5.1.

**Time Required** Generally, the same periods as given in Sect. 5.5.1 can be assumed. Small differences due to the solution behavior of the individual S/AN sample fractionated are possible in the steps Dissolution, Precipitation, Re-dissolution, and Phase separation.

**Preparatory Investigations**

See Sect. 5.5.1. Incomplete solubility in butanone may be caused by portions of higher AN content (mole fraction $> 0.7$). This can be checked by chemical analysis of the insoluble part.

**Fractionation**

The same conditions and steps as described in Sect. 5.5.1 are applied because the same fractionation procedure is used. Of course, poly(styrene) and toluene must be substituted by the copolymer and by butanone in the described steps, respectively. The nonsolvent volumes necessary for each fraction may vary from the volumes given in Sect. 5.5.1.

**Evaluation**

See Sect. 5.5.1. CC of all fractions should be checked and compared with the overall composition of the starting polymer.

### 5.5.3 Precipitation Fractionation of Poly(styrene) with Chloroform/Ethanol by Solvent Evaporation

**Materials**

**Polymer.** Poly(styrene) (10 g, the same or a similar sample as in Sect. 5.5.1)

**Solvent.** Chloroform (analytical grade), 1 l for starting polymer, 100 ml/1 g polymer fraction for re-dissolution of gel phases

**Nonsolvents.** Ethanol (96%), about 500 ml to reach the precipitation threshold; Methanol (analytical grade), 1 l/1 g polymer fraction for re-precipitation

**Others.** Inert gas for distillation under low pressure;
Cooling medium (ice/methanol mixture, condensed nitrogen), about 1 l

**Equipment**

Most devices are the same as those used in the nonsolvent-addition method. The following glassware and apparatus are additionally necessary (or useful):

- Distillation device
- Capillary tube (for inert-gas inlet)
- Graduated cooling vessel for condensation of the evaporated solvent
- Dewar flask
- Inert-gas source
- Vacuum source and control unit (including protection vessels).

A volume of 3 l of the fractionation flask is sufficient for 10 g PS to be fractionated.

**Time Required**

Generally, the time periods from Sect. 5.5.1 are valid. However, the precipitation time is replaced by the time necessary for evaporation. These periods are comparable when the distillation is carried out under low pressure. Otherwise, a longer time is necessary because of heating and cooling before and after distillation.

The time of re-dissolution and cooling of the gel phase is needless in the case of distillation at elevated temperature.

**Preparatory Investigations** See Expl. 5.5.1.

**Fractionation** Fractionation temperature: 30 °C

- Dissolution of 10 g PS in 1 l chloroform is performed as described in Sect. 5.5.1.
- Setting up of the fractionation device must include the arrangement of the distillation apparatus.
- After thermostatting, nonsolvent is added from dropping funnel only up to the first meagre turbidity (according to the cloud point determined in preparatory investigations).
- Evaporation of solvent can be carried out (i) at elevated temperature or (ii) at 30 °C under low pressure:

(i) Elevated temperature:
The dropping funnel and reflux condenser are replaced by a distillation device and stopper. The system is heated to 60–65 °C with stirring; chloroform evaporates and is condensed to measure its volume. The system is cooled after the desired volume of chloroform has been distilled (about 40 ml for the first fraction and 170 ml for the next to the last; the last fraction is obtained by a rotational-type evaporator). The gel phase appears during cooling (> 30 °C) which should occur with stirring. When the fractionation temperature is reached again, further stirring, separation, isolation, and processing of fractions as in Sect. 5.5.1.

(ii) Fractionation temperature and low pressure:
Dropping funnel, reflux condenser, and stirrer are replaced by a vacuum-distillation device, capillary tube, and stopper. The system is slowly and carefully evacuated, after connection of capillary tube and inert-gas source, so that the chloroform evaporates. (Take note of the precautions for vacuum and inert gas!) The evaporated chloroform is condensed in a cooling vessel inside a Dewar flask which allows estimation of the condensed volume. Volumes for the individual fractions correspond to variant (i). Evaporation is stopped by allowing the pressure to return to normal. The gel phase is now re-dissolved at elevated temperature, cooled, separated, isolated, and processed as described in Sect. 5.5.1.

The number of fractions and the evaporated solvent volume in each fractionation step must be estimated in connection with the masses of the foregoing (dried) fractions.

**Evaluation** See Sect. 5.5.1. The precipitation curve can be obtained by plotting the cumulative mass of fractions vs. the volume of evaporated solvent.

### 5.5.4 Precipitation Fractionation of Poly(ethylene) in Decahydronaphthalene/Dibutyl phthalate by Cooling

**Materials**

**Polymer.** Poly(ethylene) (10 g) of low (LDPE) or high density (HDPE)

**Solvents.** Decahydronaphthalene (DHN), 400 ml for the starting polymer; Xylene, 100 ml/1 g polymer fraction for re-dissolution of gel phases

**Nonsolvents.** Dibutyl phthalate (DBP), 600 ml for the starting polymer; Methanol (superpure), 1 l/1 g polymer fraction for re-precipitation; Methanol or acetone for Soxhlet extraction

**Stabilizer.** (e.g., phenyl-$\beta$-naphthyl amine), about 0.1 g

**Equipment**

- Thermostat (oil bath) equipped with glass window, stirrer, and temperature-control unit (about 20–160 °C) or, alternatively, fractionation vessel with thermo-jacket and recirculating pump
- Three-neck vessel (2 l) as shown in Fig. 5.1a–d
- Tight-fitting stirrer
- Thermometer (about 0–200 °C)
- Inlet tube for nitrogen
- Capillary tube with ground-glass joint (to function as a stopper)
- Ground-glass stoppers
- Pipettes or syringes
- Flask (1 l, thermostatted) or a second fractionation vessel to take up sol phases
- Conical flask (about 200 ml, for 1 g polymer fraction) with ground-glass joint and stopper to take up the re-dissolved gel fraction
- Soxhlet extractor
- Other devices as in Sect. 5.5.1 (without rotational-type evaporator).

**Time Required**

- Thermostatting 1 h
- Dissolution of starting polymer 2–3 h
- Cooling (depending on temperature difference) 1–3 h
- Stirring at constant temperature 0.5 h
- Phase separation overnight
- Phase isolation 1 h

Except the dissolution of the starting polymer, all periods are valid for each fraction. Time necessary for processing of fractions is not included.

**Preparatory Investigations**

- Checking for complete solubility at elevated temperature and determination of the cloud-point temperature by slow cooling
- Determination of the temperature of initial solubility by stepwise heating (e.g., 5 K) and checking the clear supernatant with excess methanol (perhaps, cooling is sufficient to give turbidity)

- The interval found should be divided into temperature steps of increasing size (e.g., 3, 5, 9, 16,. . . K) depending on the number of fractions.

**Fractionation**

- Setting up of the fractionation device: fractionation vessel (inside the thermostat or with thermo-jacket) equipped with a tight-fitting stirrer and nitrogen inlet/outlet
- Heating of 400 ml DHN to about 150–160 °C, addition of the stabilizer (with stirring)
- Dissolution of 10 g PE into the hot solvent with stirring and passing nitrogen through the apparatus
- Slow addition of 600 ml hot DBP under stirring; the mixture must keep clear. The needful dissolution temperature may be a few Kelvin higher for HDPE in comparison to LDPE.
- Exchange of nitrogen inlet and outlet for thermometer and capillary tube; slow cooling under stirring. For the first fraction, the temperature should be decreased about 3 K below the cloud-point temperature (cf. Prep. Investigations). The following temperature intervals should be extended successively.
- Stirring for about 30 minutes at constant temperature.
- Separation of the phases at constant temperature. Owing to the density of PE, the gel phase is here the upper phase.
- Isolation of the gel phase can be performed (i) by sucking off or skimming off the upper phase into the conical flask, or (ii) by draining the lower phase using a vessel as shown in Fig. 5.1d (advantageously with a thermo-jacket). In variant (i), the polymer can precipitate inside the cooler syringes. The sol phase in variant (ii) should flow into a second thermostatted vessel.
- Cooling of the sol phase as above in order to obtain the next fraction. If the sol phase became temporarily turbid (by temperature fluctuations), it should be heated before further cooling.
- Re-dissolution of the gel phase in the conical flask (and in the syringes!) or the residue in the fractionation vessel with xylene at about 120 °C and adding drop by drop to the tenfold volume of methanol with stirring.
- Soxhlet extraction of the filtered fraction with acetone or methanol, drying under vacuum at about 50 °C.
- The last sol phase can be cooled to room temperature and directly precipitated into excess methanol yielding the last fraction.

The number of fractions and the temperature intervals in each fractionation step must be estimated in connection with the masses of the foregoing (dried) fractions.

**Evaluation**

See Sect. 5.5.1. The precipitation curve can be obtained by plotting the cumulative mass of fractions vs. precipitation temperatures. The MWD of polyolefines often follows the logarithmic normal distribution (cf. Sect. 4.3.3 and Table A 7). Precipitation fractionations of PE can be influenced by both MW and branching.

## References


1. Bello A, Barrales-Rienda JM, Guzman GM (1989) In: Brandrup J, Immergut EH (eds) Polymer handbook, 3rd edn. Wiley, New York, p VII/233
2. Cantow MJR (1967) In: Cantow MJR (ed) Polymer fractionation. Academic, New York, p 461
3. Kotera A (1967) In: Cantow MJR (ed) Polymer fractionation. Academic , New York, p 43
4. Kamide K (1977) In: Tung LH (ed) Fractionation of synthetic polymers. Marcel Dekker, New York, p 103
5. Schulz GV, Dinglinger A (1939) Z Phys Chem B 43: 47
6. Flory PJ (1943) J Amer Chem Soc 65: 372
7. Meffroy-Biget AM (1955) Compt Rend 240: 1707
8. Schröder E, Müller G, Arndt KF (1988) Polymer characterization. Hanser, Munich; (1989) Polymer characterization. Akademie-Verlag, Berlin
9. Giesekus H (1967) In: Cantow MJR (ed) Polymer fractionation. Academic, New York, p 191
10. Urwin JR (1972) In: Huglin MB (ed) Light scattering from polymer solutions. Academic Press, London, p 789
11. Elias HG (1977) In: Tung LH (ed) Fractionation of synthetic polymers. Marcel Dekker, New York, p 345
12. Glöckner G (1965) Z Phys Chem 229: 98
13. Bohossian T, Delmas G (1992) J Polym Sci B, Polym Phys 30: 993

# 6 Extraction Fractionation

The reverse of precipitation fractionation can be named as "solution fractionation". This term covers various fractionation procedures, all based on the fact that a polymer species, according to its MW, is transferred from a concentrated phase into a more dilute one. This fractionation step can be modified as extraction, elution, or partition depending on the conditions of the fractionation. Although no sharp demarcations exist between these terms, the next sections shall be grouped according to the common names of the fractionation procedures. Let us now consider different extraction-fractionation procedures in this section.

## 6.1 Principles and Limitations of Application

In extraction fractionation, polymer species with lower MW are extracted from the concentrated polymer phase (gel phase) into the sol phase by increase of the dissolution power of the solvent. Thus, the fractionation starts with lowest MW in the first fraction.

As stated in Sect. 3.2, low-MW tailing is less marked in extraction than in precipitation. Fractionation in the high-MW range requires very small differences in $\chi$ between fractionation steps to obtain sharp fractions.

The most important condition for a successful fractionation is the fast adjustment of the dissolution equilibrium for the polymer species. Owing to the diffusion control of this process, a large interface between the phases is necessary. This can be achieved in various ways:

- Large polymer surface due to small polymer particles (*direct extraction* [3]) or by coating a support material with a thin polymer film (*film* and *column extraction* [3, 4]), or
- Intermixing of two liquid (gel and sol) phases (*coacervate extraction*) [1, 3].

The period of time necessary for the equilibration of the system can vary greatly and should be determined experimentally (see below).

The distribution of the polymer species between the phases is controlled by Eqs. (3.8) through (3.11).

There are many variations of extraction fractionation [1–3]. Some basic procedures are compiled in Table 6.1. Let us generally discuss these variants.

**Table 6.1.** Variants of extraction-fractionation procedures

| Procedure[a] | Direct extraction | Film extraction | Column extraction | CFE[d] | Coacervate extraction | CPF[e] |
|---|---|---|---|---|---|---|
| *Polymer*[b] | | | | | | |
| Swollen gel | + | + | + | + | | |
| Coacervate | | | | | + | + |
| *Support* | | | | | | |
| No | + | | | | + | + |
| Yes | | + | + | + | | |
| *Mode of procedure* | | | | | | |
| Stepwise | + | + | + | | + | |
| Continuous | | | | + | | + |
| Counter-current | | | | + | (+) | + |
| Column | | | + | | | + |
| Analytical | | + | + | | | |
| Preparative | | | (+) | + | + | + |
| Prefractionation | + | (+) | | | + | |
| *Principal separation mechanism* | | | | | | |
| Extraction | + | + | + | + | + | + |
| Chromatography[c] | | | + | + | | + |

[a] Procedures except CFE and CPF can be performed in two variants: (i) Constant temperature and stepwise variation of the solvent composition, and (ii) constant solvent composition and stepwise variation of temperature.
[b] State of the polymer during fractionation steps.
[c] The starting polymer is contained either in the stationary phase (column extraction) or in the mobil (CFE and CPF).
[d] Continuous film extraction – see Sect. 6.6.1.
[e] Continuous polymer fractionation – see Sect. 6.6.2.

**Direct Extraction**

In direct extraction [3], small polymer particles are brought into direct contact with the solvent to extract low-MW species from the swollen polymer. Subsequent fractions can be obtained either by different solvent mixtures of increasing dissolution power or by rising of the temperature step by step with the same solvent. Each fraction can be extracted in one step using a fractionation flask similar to Fig. 6.1, or in a multistep technique by use of a Soxhlet extractor (see below).

**Advantages**
- Simple equipment (see Sect. 6.2.1)
- Small amounts of solvent
- Direct use of the polymer.

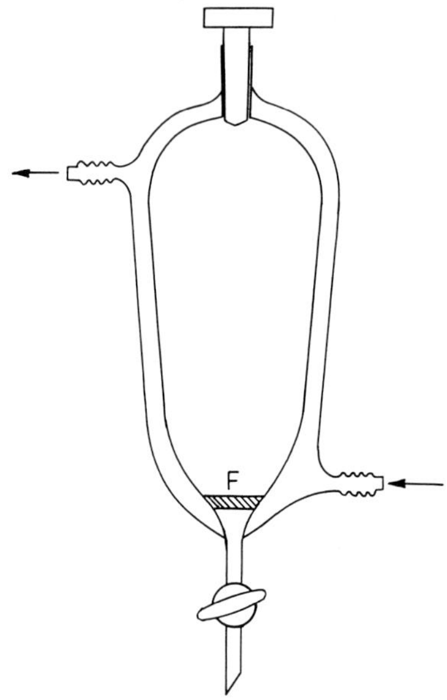


**Fig. 6.1.** Fractionation flask for direct or film extraction. *F* – glass frit; common volume: 0.5 to 2 l

**Disadvantages**
- Slow equilibration
- High swelling of the polymer particles with the risk of clogging.

Due to these disadvantages, direct extraction has only seldom been used. It can be, however, a tool for carrying out a pre-fractionation, e.g., for the separation of low-MW fractions. Extensive use has been made in the fractionation of stereospecific polymers whereby the samples were fractionated primarily with respect to tacticity (see Ref. [3]). Pre-fractionations with this method can be done on a preparative scale.

**Film Extraction** The film extraction procedure [3, 4] was developed by Fuchs [4] and is generally similar to the direct-extraction method. It works by using support material to obtain a very thin polymer film with a large surface area. As in direct extraction, the Soxhlet apparatus as well as the fractionation flask shown in Fig. 6.1 can be used (see below).

**Advantages**
- Attainment of the equilibrium within a reasonable time scale, i.e., short fractionation time.
- Simple to carry out and only small amounts of solvents are needed.

**Disadvantages**
- Drying of the film requires experience. Both the adhesion of the polymer to the support material and the rate of equilibration are influenced by the degree of drying.

- After some extraction steps the polymer film often detaches itself from the support.

The last problem which depends on the adsorption strength between support and polymer and increases with increasing MW seems to be the reason why this fractionation procedure can only be successfully applied to a few polymers. Owing to the very thin polymer film the method has usually been used on an analytical scale. However, Blair [5] has developed and used a continuous film-extraction method for preparative fractionations (see Sect. 6.6.1).

Let us come back to the two possibilities of "flask extraction" and Soxhlet extraction:

When one extracts each fraction in one step with successively increasing solubility strength (expressed by descending values of $k$, cf. Eq. (3.8)), Eqs. (3.10) and (3.11) are valid for the gel and sol phases, respectively (for details see Table A 8 in the Appendix).

In Soxhlet extraction, the polymer is brought into contact several times with the same extracting agent working with the same $k$ value and nearly the same volume ratio (see Table A 8).

Comparison of the related fractions ($w_{i,1}'$ and $\sum w_{i,1x}'$, or ($w_{i,2}'$ and $\sum w_{i,2x}'$, etc., cf. Table A 8) shows that the fractionation effect is different: More of the polymer species $P_i$ is, using the same values for $k$ and $V''/V'$, extracted using the Soxhlet procedure than by the "flask extraction". In other words, certain kinds of molecules can be extracted more completely from the swollen gel phase by the Soxhlet procedure. However, this fraction will be contaminated simultaneously with higher-MW portions of the polymer. Thus, Soxhlet extraction does not lead generally to narrower fractions. Therefore, it should be applied chiefly for pre-fractionation to eliminate almost completely low-MW or, e.g., atactic portions of a polymer.

**Column Extraction**

The problem of detachment of the polymer in film extraction can be largely eliminated by use of a polymer-coated support material (e.g., small glass beads or sea sand) in a column which will subsequently have solvent passed through, either in an upward or downward direction [2, 3, 6]. The extraction is performed in steps by the use of different solvent mixtures or by different extraction temperatures. Column procedures with continuous gradients will be described in Sects. 7 and 8. Many aspects of the stepwise extraction play a role likewise in gradient techniques.

The separation mechanism of column extraction includes chromatographic steps because the gel phase at different loci in the column comes into contact with sol phases of different compositions caused by the increase of the polymer concentration in the sol phase along the column. In addition, the starting zone of the polymer-coated support is always extracted by fresh solvent (and, likewise, a specific locus inside the packing is repeatedly extracted by a sol phase of nearly constant composition). Thus, working conditions are similar to those in Soxhlet extraction.

**Advantages**
- Automation is possible.
- Large-scale fractionations can be done.
- Adjustment of the equilibrium is better than in direct or film extraction.
- Detachment of the polymer is hardly possible because the coating is fixed better on the particles of the support which themselves are firmly fixed mechanically in the column.
- Chromatographic steps improve the fractionation efficiency.

**Disadvantages**
- The method requires more expenditure and preparatory work than direct extraction or film extraction.
- Bigger amounts of solvent are necessary; high diluted solutions result as fractions.
- Column blocking might occur if detachment or swelling of the polymer exceed a certain limit.

**Coacervate Extraction**

Let us consider in coacervate extraction [1, 3] only cases in which a liquid, polymer-rich gel phase (coacervate) is extracted by a solvent forming a sol phase. Both the sol phase (or extracting agent) and the gel phase (coacervate) contain the same solvent and nonsolvent and are only formed together with the polymer fractionated (i.e., solvent and nonsolvent are completely miscible without the polymer). Thus, this procedure is a liquid–liquid extraction. Subsequent extracting agents may have different compositions at constant temperature or constant compositions usually working at rising temperatures.

Other, seemingly very similar procedures in which the polymer is distributed between two immiscible liquids shall be discussed under the term "partition" in Sect. 11.

The general principle of a coacervate extraction is as follows: At first, a liquid gel phase containing almost the total amount of polymer is produced from the polymer solution by addition of an appropriate volume of a nonsolvent at a certain temperature. Then, the liquid phases are intermixed to ensure the extraction equilibrium. When the droplets are aggregated into bulky phases, the diluted one, which contains the first polymer fraction, is isolated. Then, the next volume of extracting agent with higher solvent content (constant temperature) or higher temperature (constant composition) is used to obtain the next fraction, etc. The procedure resembles "upward precipitation fractionation" (See pp. 45, 47, and 51), but, unlike the latter, the gel phase is not formed anew in each fractionation step of the coacervate extraction.

When a sharp fraction has to be obtained on preparative scale by one-step extraction, a considerable mass of the starting polymer [1] and great volumes of solvent and nonsolvent are required. Therefore, multiple extraction via a *counter-current* procedure (See Sects. 6.5.2 and 6.6.2) is preferred, which may yield sharp fractions [1, 7].

**Advantages**
- Rapid adjustment of the equilibrium occurs.
- Suppression of the low-MW tailing (cf. Sect. 3.2) is successful.
- Simple to carry out.
- Compared with "upward precipitation fractionation", smaller volumes of solvent and nonsolvent are necessary.
- Automation of counter-current procedures is possible.

**Disadvantages**
- The coacervate must remain a liquid, i.e., flocculation of the polymer in the concentrated phase must not occur.
- The estimation of the extracted polymer in the sol phase is difficult.

Coacervate extraction has mostly been used for the fractionation of polyolefines [3] and to some extent also of polyesters and polyamides [8].

In recent years, new developments in coacervate extraction have taken place leading to "continuous polymer fractionation" which is a chromatographic counter-current process on the preparative scale (see Sect. 6.6.2).

## 6.2 Direct Extraction

### 6.2.1 Equipment and Materials

The implementation of a direct extraction is possible by use of common laboratory glassware. The main item is the extraction vessel, which can be a conical flask in the simplest case, a flask as shown in Fig. 6.1, or a Soxhlet extractor. The first two vessels mentioned must allow thermostatting and shaking. A special vibratory mixer is described in Sect. 6.2.2.2.

The isolation of the sol fraction can be carried out with a syringe by hand, with a syphon similar to the one in Fig. 5.2c–e, or through the stopcock of a vessel as shown in Fig. 6.1. The stopcock should be protected from the extraction chamber by a sintered glass frit or by glass wool. In a Soxhlet apparatus, the sol fraction is separated a priori.

Further processing of the sol fractions (such as evaporation of the solvent, precipitation, re-dissolution, re-precipitation, drying) requires analogous devices as discussed in Sect. 5.2 (Table 5.1). The direct extraction requires only solvent and nonsolvent (or some different solvents) for the extracting agent of the successive fractions. At least 100 ml extracting agent are necessary for 1 g polymer when using an extraction flask. In the Soxhlet procedure, the solvent volume is determined by the volume necessary for the reliable functioning of the apparatus.

For the processing of the fractions, the same estimations as for precipitation fractionation (Sect. 5.2) are valid.

### 6.2.2 Specific Steps of Implementation

Consecutive steps of direct (DE) and film extraction (FE) including some preparatory steps are summarized in Table 6.2.

#### 6.2.2.1 Preparatory Investigations

Preparatory steps should involve

- selection of the extracting agent for extraction of the different fractions according to volume and composition
- studying of swelling behavior of the polymer
- choice of the extraction temperature
- determination of the period of extraction.

**Table 6.2.** Schematic course of the direct (DE[a]) or film extraction (FE[a])

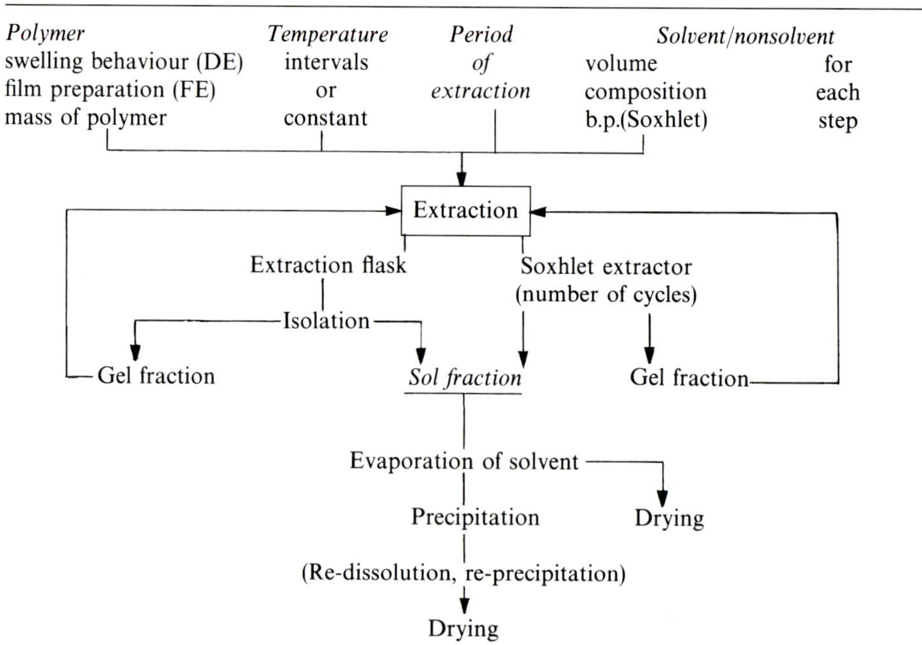


[a] Abbreviations are only used for different steps or aspects of the procedures.

**Selection of the Extracting Agent**

When the solvent/nonsolvent pair is chosen (if possible by consideration of the aspects discussed in Sect. 4.1.2 and Table 4.2), the limits of complete solubility and insolubility of the polymer must be determined. These limits can be found by turbidimetric titration (cf. Sect. 5.4.2 and Fig. 5.4). The cloud point yields the limit of solubility. The limit of complete insolubility, found by turbidimetric titration, can be checked by excess nonsolvent addition to the supernatant of a polymer/solvent/nonsolvent mixture of corresponding composition. Now, the fractionation range among these limits can be subdivided into extraction steps depending on the desired number of fractions. The mass of the extracted polymer is directed by Eqs. (3.8) through (3.11) (cf. Table A 8 in the Appendix). Both increasing dissolution power (decreasing value of $k$) and growing volume of the sol fraction (extracting agent) extend the mass of polymer in the sol fraction. The volume of the gel fraction can be supposed as approximately constant in the direct-extraction procedure. Consecutive extracting agents should be tested before the actual fractionation.

A numerical estimation of this problem, as a first guess only, is also possible. This can be outlined here only briefly by the following steps:

- Estimation of the solubility parameters of polymer and (mixed) solvent at a given temperature (cf. Sect. 3.1 and Tables A 1 through A 3)
- Calculation of the Huggins parameter $\chi$ for different compositions of the extracting agent according to Eq. (3.7)
- Estimation of $k$ in Eqs. (3.8) through (3.11) from $\chi$ (see Sect. 3.2 and Refs. [1, 9, 10])
- Calculation of the values $w_i'$ for corresponding $P_i$ using different $k$ and $V''/V'$ according to Eqs. (3.10), (3.11), and corresponding equations in Table A 8.

**Swelling Behavior of the Polymer**

Swelling behavior depends on the extracting agent, temperature, and polymer consistency (e.g., powder or flakes). Generally, the formation of a compact polymer body which obstructs the equilibration should be prevented. Preceding tests (different solvent mixtures and temperatures, kind of shaking or stirring) must be done for this purpose.

**Extraction Temperature**

The temperature should be chosen either from a practical point of view or from the aspects just discussed.

**Period of Extraction**

The extraction time must be determined experimentally. In order to do this, the mass of the polymer in the sol phase after different periods of time can be determined by differential-refractometry, densitometry, or gravimetrically. It should be noted that the period necessary for equilibration may depend upon experimental details, e.g., the intensity of mixing (see below).

#### 6.2.2.2 Fractionation Steps

**Mixing** Different procedures can be chosen for the mixing of polymer and extracting agent, e.g.,

- addition of the extracting agent to the polymer with or without shaking
- addition of the polymer in the moved extracting agent (perhaps only possible for the first fraction).

One should use the procedure which yields the best spatial distribution of the polymer inside the extraction vessel.

This step is invariable in the Soxhlet technique.

**Extraction**

- "*Flask extraction*" should be more effective with shaking than with stirring. If necessary, the polymer must be dispersed mechanically (e.g., by glass spheres inside the extraction flask) during the extraction steps.
- "*Dynamic direct extraction*" [11] can be performed in a thermostatted extraction vessel equipped with a frit and, as an essential item, a vibratory mixer. This mixer, which is distinctly more efficient than a normal stirrer, is also able to disperse highly swollen polymer gels. The equilibration time is sufficiently short.[1]
- *Soxhlet extraction* is not suitable for mixed extracting agents due to the difference in boiling points of the components. Alternatives are the use of different single solvents having increasing dissolution power, or variation of boiling points of a single solvent as a function of pressure.

The required number of extraction cycles can be checked in the following way: A cycle from the Soxhlet extractor must be separated and checked by addition of a strong precipitant. The next extraction step can be started if the turbidity is now only slight (distinctly weaker than in foregoing cycles – the extracted polymer mass does not reach zero completely, cf. Sect. 6.1 (p. 67), film extraction).

**Isolation Sol Fraction** The sol fraction can flow out through the stopcock by use of an extraction flask similar to Fig. 6.1. If the sintered glass disk becomes clogged, the sol phase should be isolated by a syringe or a syphon (cf. Fig. 5.2c–e). Short washing of the residue (short in comparison to the period of extraction) with a small volume of the extracting agent used last can be advantageous before the next extracting mixture is used.

[1] Analytical and preparative extraction fractionations of poly(ethylene) have been reported, which took less than six hours and reached efficiencies like those in SEC [11].

In Soxhlet extraction, the sol phase is isolated a priori at the bottom of the apparatus.

**Processing**

The sol fraction can be processed after isolation by evaporation or precipitation (as described for precipitation fractionation, see Sect. 5.3.1 and also Sects. 4.2.2 and 4.2.3).

## 6.3 Film Extraction

### 6.3.1 Equipment and Materials

The same equipment as in the direct-extraction procedure can be generally used (cf. Sect. 6.2.1). Additionally, some extra glassware is required to prepare the polymer film. Size and shape of the necessary vessels (flasks, basins etc.) depend on the size, shape, and material of the support and on the mass of the polymer (see Sect. 6.3.2.2). The polymer film must be dried. This can be done using an exhauster and ventilator at room or slightly elevated temperature – care must be taken to prevent contamination of the air with toxic solvent vapors.

The following materials are necessary

- Solvent for the coating procedure (about 100 ml per gram polymer)
- Support material, mostly aluminum or textiles (1000 to 2000 $cm^2$ per gram polymer [4])
- Solvents and nonsolvents as for the direct-extraction procedure.

### 6.3.2 Specified Steps of Implementation

The course of the film extraction follows the scheme given in Table 6.2.

#### 6.3.2.1 Preparatory Investigations

Selection of the mixtures and of the volumes of the extracting agents, choice of the extraction temperature and of the period of extraction can be carried out analogously to the direct extraction. A shorter period of extraction, necessary for adjustment of the extraction equilibrium, should be found in the film-extraction procedure. This period of time increases with growing MW and must be longer for associating polymers [4(1951)].

Special investigations are required for the *film preparation.* The adhesive power of the polymer film may depend on various conditions:

- Kind of support and property, quality, and cleanliness of its surface
- Concentration and solvent of the polymer solution
- Thickness of the polymer film
- Degree of drying.

Optimal conditions of the film preparation must be found experimentally. For details see the next section.

#### 6.3.2.2 Fractionation Steps

**Preparation of the Polymer Film**

The polymer must be dissolved in a *concentration* range from 1 to 10 percent. The chosen concentration depends on the viscosity of the solution, which is influenced by the MW of the polymer and by the solvent used. The choice of the solvent should be made considering the boiling point. When the solvent evaporation is too fast then the film adhesion may be insufficient. The solvent used should be preferably one of the components of the extracting mixture.

The *thickness* of the film can be varied by the viscosity (concentration) of the solution and by repeated dipping of the support into the solution. It must be apprehended that growing film thickness increases the chances of dislodging.

The *adhesive power* of an aluminum foil can be enhanced when the surface is mechanically roughed. Contamination of the surface reduces the adhesion. In addition, the adhesive power depends on the polarity of the polymer. For instance, PS is less suitable for film extraction because of its poor adsorption power whereas poly(vinyl acetate) (used by Fuchs [4]) shows a strong adherence.

*Drying* of the polymer film must be performed carefully and sufficiently slowly. Drying in a vacuum cannot be recommended because of the increasing likelihood of dislodging the polymer.

The *load* of the support with the polymer can be controlled by weights of the unloaded and loaded support. Note, that usually both sides of the support are coated. Customarily, loads from 0.5 to 1 mg/cm$^2$ are used denoting a thickness of about 10 μm.

Finally, the coated support is cut up, coiled, or crumpled and placed in the extraction vessel.

**Extraction**

- "*Flask extraction*": After addition of the extracting agent (having the extraction temperature) the vessel should be shaken sightly. Violent shaking promotes the detachment of the polymer. Likewise, scrubbing off of the polymer must be prevented. Very long contact time increases the risk of the film being dislodged. Fractionation flasks with a glass frit (Fig. 6.1) are to be preferred because dislodged particles are held back.
- *Soxhlet extraction* is analogous to direct extraction.

**Isolation and Processing**

Isolation and processing of the sol fractions: See direct extraction (Sect. 6.2.2.2).

## 6.4 Column Extraction

### 6.4.1 Equipment and Materials

In column extraction one needs apparatus for the following basic procedures [2, 3, 6, 12]:

- Loading the support material
- Extraction
- Collection of the fractions
- Processing of the fractions.

**Loading the Support Material**

If the support is coated with the polymer sample outside the fractionation column, the following apparatus is required:

- Glass beaker having a volume greater than the volume of the support material
- Glass stick (or similar device) for agitation of the polymer/solvent/support mixture
- Ventilator and exhauster.

Techniques for coating the support within the column will be described in Sect. 6.4.2.

**Extraction**

The apparatus for the column-extraction procedure is shown schematically in Fig. 6.2. Usually, glass columns furnished with a jacket for thermostatting are used. The advantage is their transparency, which allows one to inspect the

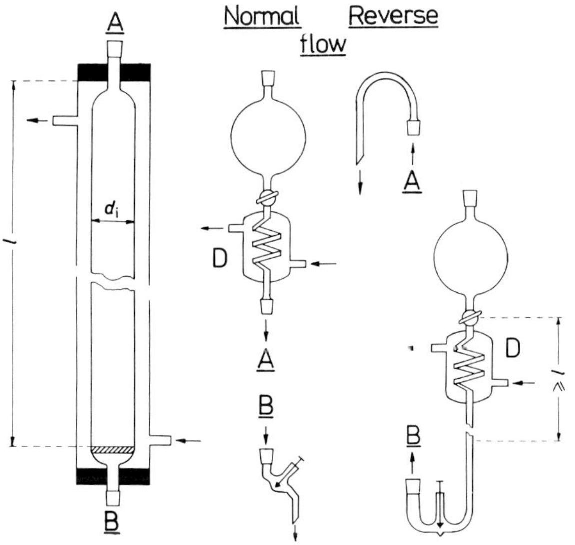


**Fig. 6.2.** Simple equipment for column-fractionation technique. A, B – variants for normal and reverse flow, respectively; **D** – device for degassing of the (mixed) solvent. Common column parameters are: $d_i$ = 2 to 7 cm, $l$ = 35 to 150 cm

column bed during extraction with respect to channeling or gas bubbles (see below). Metallic columns (aluminum, stainless steel) are recommended for higher temperatures (> 130 °C) [3].

The column must be connected with a reservoir for the extracting agent, completed by devices for degassing and thermostatting. A finely-adjustable outlet allows one to control the flow rate of the extracting agent. Simple variants for normal and reverse flow, using only gravity, are shown in Fig. 6.2. The bottom of the column should contain a sintered glass frit or should be plugged with glass wool to fix the support material. The column jacket is fitted with a thermostat.

Columns with the parameters given in Fig. 6.2 are mostly used for analytical or semi-preparative fractionations. Large-scale stepwise column extractions of a swollen gel phase can be performed similarly to gradient-elution or Baker-Williams fractionation (see Sects. 7 and 8) and will be discussed in Sect. 9.2.

**Collection and Processing of the Fractions**

According to the number and volume of the fractions, relative collecting vessels are necessary. A fraction collector may be used.

Similar devices to those for precipitation fractionation (See Sect. 5.2) and direct extraction are required for the processing of the fractions.

The following are descriptions of the necessary materials:

**Support**

The mass or volume of the support depends upon the volume of the column. Polymer/support ratios will be discussed in Sect. 6.4.2. Sea sand and glass beads with diameters of about 0.2 to 0.5 mm are mostly used as supports. Note, that glass beads contain alkalies which can lead to complications with respect to chemical stability of some polymers. Metal powder can also be used as support, but, attention must be paid to potentially catalytic influences on the polymer.

**Solvent for Loading**

This solvent should be a relatively poor one but highly volatile. About 50 ml solvent per gram polymer are necessary.

**Nonsolvent for Extracting Mixtures**

Usually, one needs a nonsolvent-rich mixture to fill the column before the fractionation starts, i.e., this mixture must have a composition in which the polymer is insoluble. The volume of the mixture is the difference between the column volume and volume of the support in the column (interstitial volume, see Glossary). Additionally, the nonsolvent is one component of the extracting agents. Its essential volume depends on the volume and composition of the extracting mixtures.

**Solvent for Extracting Mixtures**

A solvent is required in all extracting agents (increasing with growing fraction number). The volume and composition of the solvent/nonsolvent mixtures (extracting agent) depend upon the following parameters:

- Mass of the polymer to be fractionated
- Number of fractions (or, in other words, differences in the dissolution power of successive extracting agents)
- Heterogeneity of the polymer
- Column volume
- Flow rate of the extracting agent.

Increase of these parameters, especially a flow rate which is too high and small composition steps of extracting agents, leads to an enlargement of the total volume of the solvent/nonsolvent mixtures. Additional to the materials discussed here, solvent and nonsolvent for precipitation, re-dissolution and re-precipitation of the fractions may be necessary (cf. Sects. 4.2.2 and 5.2).

### 6.4.2 Specific Steps of Implementation

Before a column extraction can start, similar points to those given at the top of Table 6.2 must be checked.

**Extracting Agents and Temperature**

The choice of solvents and nonsolvents should take into account the topics discussed in Sect. 4.1.2. Determination of composition and temperature of the different extracting mixtures can be performed as for the direct extraction (p.71). The steps in the solvent/nonsolvent ratio of the extracting agents or the temperature steps with constant solvent/nonsolvent ratio must be the smaller the higher the MW range of the polymer is. This can be well implemented by use of a poor solvent and a weak nonsolvent. Required volumes depend upon the parameters given in the preceding section.

**Cleaning the Support**

The most commonly used glass beads or sea sand should be washed with concentrated hydrochloric acid, and then repeatedly with distilled water, and (after drying) with the solvent used in the fractionation. Sea sand should be annealed before the cleaning procedure.

**Loading the Support**

The support can be coated by the polymer outside as well as inside the column [3, 6, 12]. For preparative fractionation, the procedure outside the column is recommended.

- Loading *outside* the column: The polymer is dissolved in a highly-volatile solvent with a concentration of usually one to five percent. Dichlorometh-ane or butanone are often used as solvents. The solution is mixed with the support, afterwards, the solvent is evaporated slowly, e.g., by use of a heated air stream. Stirring is necessary to avoid the formation of clumps. Small clumped parts of the polymer/support mixture can be *carefully* mortared. Coating the support should be combined with a fractional precipitation of the polymer on the support, i.e., the highest-MW parts should be precipitated first and the lowest-MW molecules last. This is important especially for an effective fractionation of polymers having a narrow MWD and can be achieved by use of poor solvents or of suitable solvent/nonsolvent mixtures (boiling points of the components!).

Another variant is the dissolution of the polymer in a (mixed) solvent at elevated temperature and fractional precipitation by cooling. Preliminary tests are necessary for this procedure.

- Loading *inside* the column: This technique is only possible if precipitation by cooling can be used. The polymer dissolved in a poor solvent or a solvent/nonsolvent mixture at elevated temperature is filled in the column (interstitial volume) and afterwards the support is added. The equal, fractional precipitation can be reached by slow cooling.

The load of the support, i.e., the ratio polymer/support varies between wide limits. Examples in the literature (see Refs. [3, 6, 12]) give values from about 0.025 mg/cm$^2$ to more than 6 mg/cm$^2$. We have achieved good results with 0.2 to 0.3 mg/cm$^2$. The last values correspond, for glass beads with a diameter of 0.4 mm, to mass ratios of 12 to 16 mg polymer per gram support (or ca. 22 to 30 mg per cm$^3$ bulk volume of the support). The ratio should be the smaller the narrower is the MWD, e.g., 2 mg/g for narrow MWD and 8 mg/g for broader MWD are recommended [6].

**Design of the Column**

A schematic picture of the column is given in Fig. 6.2. Ground-glass joints can be sealed using poly(ethylene) or teflon gaskets. Both downward and upward directions of flow are possible. The upward direction should be preferred if there is a danger of the extraction bed or the glass frit being clogged by dislodged polymer particles. With upward flow, the top of the support should be covered by a porous disk or by a sieve to stabilize the support.

All electrical and heating devices must be positioned so that no danger (especially fire) can arise from solvents or nonsolvents. Likewise, care must be taken to avoid contaminating the air.

**Filling the Column**

The filled column must be free of gas bubbles and channels in the support. This can be achieved by the following procedure: The empty column is filled with the (mixed) nonsolvent relative to the volume of the liquid in the filled column (i.e., the interstitial volume). The coated support is now added to this nonsolvent and a homogeneous and gas-free column bed is obtained.

**Thermostatting**

The best variant of thermostatting is the use of a jacket through which a liquid at a constant temperature is passed. If the extraction steps are obtained at different temperatures, a second thermostat working at the next higher temperature and also connected to the jacket by taps is useful. In this way an immediate temperature change is possible.

**Degassing of the Solvent**

Gas bubbles in the column bed disturb the steady flow of the extracting agent. Therefore, both the starting mixture of the nonsolvent (cf. filling the column) and the following solvent/nonsolvent mixtures must be degassed before they pass the column. This can be usually achieved by heating above the fractionation temperature before use or continuously in a special device of the fractionation apparatus (See Fig. 6.2). Ultrasonic treatment is also a tool for degassing.

Parameters influencing the fractionation efficiency are

**Parameters of the Extraction Procedure**

- height of the extraction steps (difference in composition or temperature of the extracting agents – see above)
- temperature constancy
- flow rate
- extraction period of the fractions (volume of extracting agent per fraction)
- load.

Let us discuss these parameters except the first one (see above): *Constant temperature* is very important and should be given more attention than the absolute extraction temperature. The latter should not be higher than necessary.

The *flow rate* of the extracting agents must be optimized experimentally. The commonly used flow rates, given in the literature and related to the cross-section of the column, cover a range from about 1 to 30 ml/h cm$^2$. No serious influence on fractionation efficiency is expected from the flow rate in the column procedure discussed here. The *extraction period* depends upon the mass of the fraction, the interval of the extraction power, the width of MWD, and the flow rate (which influences the degree of the extraction equilibrium in the swollen gel phase). The extraction period grows with increases in the parameters mentioned. The extraction of a certain fraction is finished when a droplet of the extract does not cause any turbidity in a strong but miscible nonsolvent.

The *load* was discussed above. When the whole support material is coated, the total mass of the polymer can reach a few grams. Overload may produce clogging of the column, reverse-order fractionation, and diminution of the fractionation efficiency.

The extraction can be performed (related to a fraction) continuously or in the stop-and-go variant. The latter enables a better extraction equilibrium and requires less extracting agent (eluent) – see Sect. 6.7.2.

**Processing of the Fractions**

Owing to the high dilution of the fractions, a rotational-type evaporator should be used before further processing which can be performed as discussed in Sects. 4.2.2, 4.2.3 and 5.3.1.

**Purging the Column** Column and support must be purged together with or after the last fraction. A sufficient volume (at least twice the liquid volume inside the column) of a strong solvent should be used. Alternatively, one lets a sufficient volume of the last extracting agent flow through the volume at an elevated temperature. It is possible to check the procedure by seeing if precipitation of the polymer in a strong nonsolvent occurs (turbidity).

Problems of trouble-shooting will be discussed together with other column-fractionation procedures in Sect. 9.1.

## 6.5 Coacervate Extraction

### 6.5.1 Equipment and Materials

Equipment and materials are similar to "upward precipitation fractionation" (see Sect. 5.2). An essential item for coacervate extraction is the extraction vessel. It should have a slender design, e.g., as shown in Fig. 6.3, and can be equipped with a stirrer, a dropping-funnel, and (if necessary) a reflux-condenser. The same devices as described for precipitation fractionation can be used for thermostatting and for isolation and processing of the sol fractions (see Sect. 5.2, Table 5.1, and Fig. 5.2).

For a counter-current method of coacervate extraction (see p.69 and Sect. 6.5.2.3), a set of extraction vessels and devices for the transport of the sol phases are required. A Craig partition apparatus or the like would be advantageous.

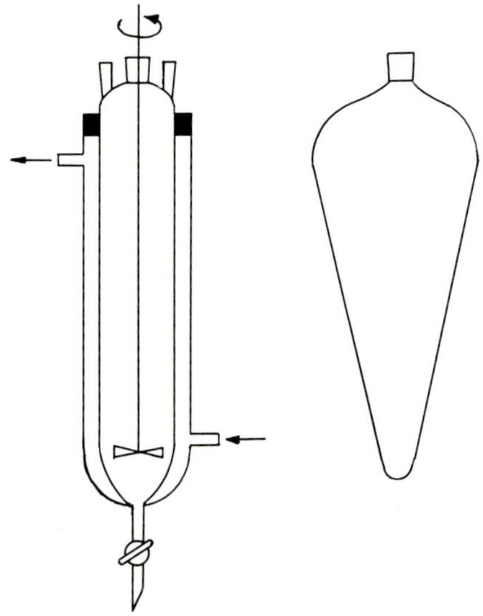

**Fig. 6.3.** Examples of fractionation vessels for coacervate extraction (common volume about 2 to 4 l). When intermixing of the phases is done by shaking instead of stirring, a thermostatting jacket is advantageous

Materials necessary for coacervate extraction are:

- Solvent for dissolution of the polymer
- Nonsolvent to produce the coacervate
- Solvent and nonsolvent for different extracting mixtures.

The ratio of polymer to solvent is similar to "upward precipitation fractionation" (see Sect. 5.2), i.e., 50 g polymer dissolved in 1 l solvent could be fractionated in a vessel of 3–4 l. More nonsolvent may be necessary for the coacervate formation in comparison to the nonsolvent required for the first precipitation (about 90% of the polymer) in "upward precipitation fractionation". The amount of solvent and nonsolvent for extraction depends on the number of fractions, on composition and volume of the extracting mixtures and can only be predicted with great difficulty. According to Eq. (3.11), the mass of polymer in the sol phase can be influenced in the coacervate extraction by composition or temperature ($k$ in Eq. (3.11)) as well as by the volume ($V''/V'$) of the extracting agent. Increase of the volume ($V'$) instead of the diminution of $k$ leads to a greater sharpness of the extracted fraction (cf. Sect. 3.2). The same estimations as for precipitation fractionation (Sect. 5.2) can be made for further processing of the fractions.

### 6.5.2 Specific Steps of Implementation

#### 6.5.2.1 Preparatory Investigations

Conditions of the coacervate formation and the gradation of the extracting mixtures must be found out experimentally. Investigations of the solubility behavior (complete coacervate formation, complete dissolution) can give hints and limits.

Experimental conditions such as

- period of extraction (which will be relatively short in this procedure)
- temperature
- composition and volume of the solvent/nonsolvent mixtures for each step

can be determined and monitored as described for the other variants of the extraction fractionation (see Sect. 6.2.2.1 and Table 6.2).

#### 6.5.2.2 Simple One-Step Coacervate Extraction

A schematic course is shown in Table 6.3.

**Dissolution, Thermostatting, and Coacervate Formation**

These steps can be performed generally as described for "upward precipitation fractionation" (see Sect. 5.3.1). Conditions for the coacervate formation (polymer concentration, temperature, solvent/nonsolvent pair) must guarantee a liquid coacervate (at least till the extraction equilibrium is reached, see below).

**Table 6.3.** Schematic course of the coacervate extraction

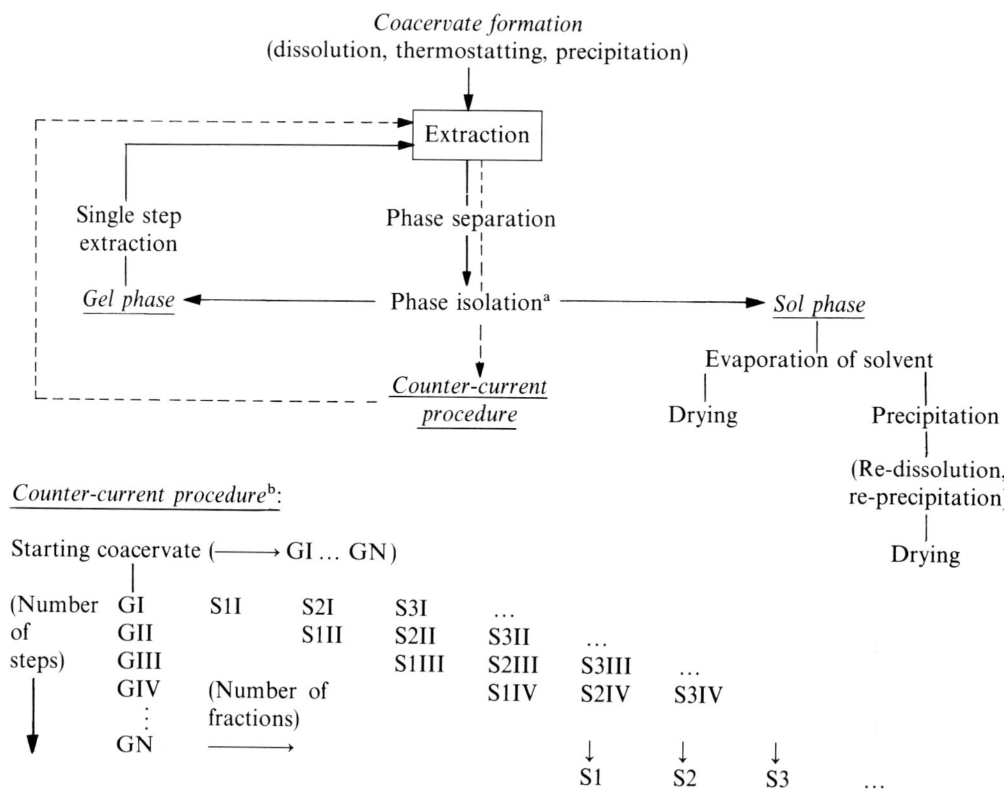


[a] This step can be omitted by use of an apparatus to carry out counter-current extraction.
[b] The coacervate is assumed to be the stationary phase (for detailed description see text). Symbols mean: G – gel phase (coacervate), S – sol phase (finally the fraction), roman numerals – number of the fractionation step (vessel), arabic numerals – number of the fraction (Different fractions are caused by different composition or temperature of the extracting agent).

The coacervate can be produced also by sufficient cooling of a homogeneous solution below the dissolution temperature of the polymer/solvent/nonsolvent system.

**Extraction** Extraction is performed with stirring or shaking for a sufficiently long period of time. Four variants of the extraction step can be discussed:

- Addition of extracting agent to the coacervate at constant temperature.
- As the first procedure but with heating of the system to reach a homogeneous solution and afterwards cooling to extraction temperature under stirring or shaking.

- Addition of the solvent portion of the extracting mixture in a first step whereby dissolution may occur. The extraction equilibrium is reached in a second step by addition of the necessary portion of the nonsolvent.
- Addition of the extracting agent of constant composition to the coacervate which was produced by cooling; enhancement of the extraction temperature. If necessary (flocculation), the temperature can be elevated temporarily above the dissolution temperature and is diminished afterwards to extraction temperature.

The best course must be checked experimentally for the given system. The second and third variants could be advantageous if the polymer in the coacervate tends to precipitate or to re-crystallize.

**Phase Separation, Isolation, and Sol-phase Processing**

The same aspects and procedures as for precipitation fractionation, especially "upward precipitation fractionation", are relevant (see Sects. 4.2.2, 4.2.3, and 5.3.1).

#### 6.5.2.3 Counter-Current Procedure

A simple variant could be realized as follows (cf. Table 6.3):

**Coacervate Formation**

After coacervate formation discussed above, the coacervate is distributed in several extraction vessels (GI–GN).

**Extraction**

Extracting agent S1 is added to vessel I (GI) leading to sol fraction S1I. The latter is transported into vessel II (GII) to obtain sol fraction S1II which is now added to vessel III (GIII), etc.

The coacervate remaining in vessel I is now treated with the next extracting agent S2 to form sol fraction S2I. This sol fraction passes on to the next vessels yielding sol fractions S2II, S2III, etc. The same is performed with the next extracting agent S3, etc. Finally, one obtains the consecutive fractions S1, S2, . . . after N counter-current steps.

**Phase Separation, Isolation, and Processing**

These steps are analogous to the one-step extraction described above.

## 6.6 Continuous Procedures of Extraction Fractionation

Continuous procedures are important for large-scale counter-current fractionations (see Table 6.1).

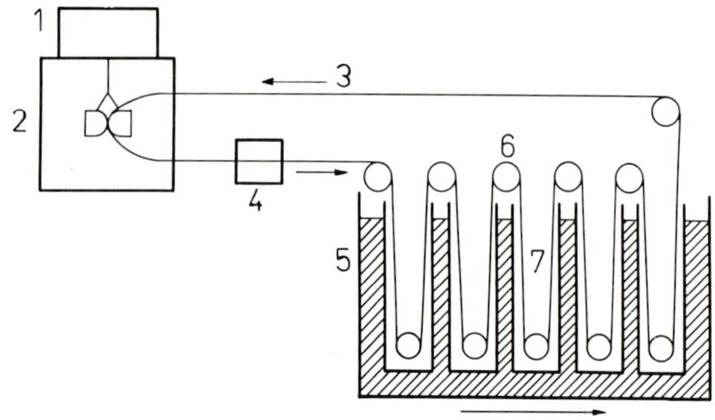


**Fig. 6.4.** Schematic representation of the continuous film-extraction apparatus by Blair. *1* – reservoir, *2* – coating device, *3* – continuous belt, *4* – drying device, *5* – constant temperature bath, *6* – sprockets, *7* – tubes with increasing solvent power (Ref. [5] D.E. Blair, J. Appl. Polym. Sci. 14 (1970). Copyright: John Wiley & Sons, Inc. Reprinted with permission of John Wiley & Sons, Inc.)

### 6.6.1 Continuous Film Extraction

This method was developed by Blair [5]. The apparatus used is schematically shown in Fig. 6.4: A continuously running, endless belt of poly(ethylene terephthalate) is coated on both sides with a viscous polymer solution, afterwards dried and finally led through glass vessels filled with extracting agents of increasing solvent power. The speed of the belt depends on the fractionation system, especially on the MW of the polymer.

Analytical fractionations can be performed in one run of the belt. Continuous, repeated coating of the belt is used for preparative fractionations. In this way, Blair fractionated the following polymer samples: 348 g Neopren W (10 fractions, 25 days), 50 g poly(tetramethylene ether glycol) (20 fractions), 61 g poly(methyl methacrylate) (11 fractions, 1.5 days), 1048 g phenol-formaldehyde resin (7 fractions, 10 days), or 271 g fluorelastomer (10 fractions, 5 days).

In Blair's method, the polymer film can be regarded as the mobile phase of a chromatographic counter-current extraction process.

### 6.6.2 Continuous Polymer Fractionation[2]

This special form of a liquid–liquid extraction is based upon theoretical considerations by Englert and Tompa [7] and has been elaborated by Wolf et al. [1, 13–18].

[2] This section was prepared using additional information kindly given by Prof. B.A. Wolf, K. Schultes, and R. Mertsch. Their help is gratefully acknowledged.

### 6.6.2.1 Principles and Limitations of Application

In the continuous polymer fractionation (CPF), the polymer dissolved in a solvent/nonsolvent mixture (called the feed) is treated with a liquid extracting agent which contains the same kind of solvent and nonsolvent with a higher content of nonsolvent than the feed. The treatment forms the coacervate due to the limited solubility of the polymer in the solvent/nonsolvent system. The coacervate is continuously extracted inside a counter-current column by the liquid extracting agent whereby the polymer molecules will be distributed over the counter-current phases according to their MW. Finally, the feed leaves the column as a polymer-rich, liquid gel phase, and the originally polymer-free extracting agent flows off as a sol phase containing the low-MW part of the polymer. Thus, the original polymer can be split into two fractions at a desired MW in one continuous run.

The position of the cut through a MWD has to be decided first and leads to the ratio $\dot{G}$ of the fluxes of mass of polymer, $\dot{m}_P$, in sol and gel phase, $\dot{m}'_P$ and $\dot{m}''_P$, respectively:

$$\dot{G} = \dot{m}'_P / \dot{m}''_P \tag{6.1}$$

If more than two fractions are to be produced in more than one fractionation run, $\dot{G}$ should be chosen so that the gel phase n can be used as feed n + 1 in the next run, whereas the sol phases can be used directly as fractions (i.e., $\dot{G} < 0.5$ in the first run and $\dot{G} \approx 1$ in the last). The possible number of fractions depends on the mass of polymer to be fractionated and on the dimensions of the column (see next section).

The $\dot{G}$ value is connected with the "working point" which is the overall composition in the entire column under stationary operating conditions formed by the feed (including the polymer) and extracting agent according to their flow rates $\dot{V}$:

$$w_i^{WP} = (\dot{V}^{EA} \cdot w_i^{EA} + \dot{V}^{FD} \cdot w_i^{FD}) / (\dot{V}^{EA} + \dot{V}^{FD}) \tag{6.2}$$

Here, $w_i$ represents the weight fraction of solvent, nonsolvent, and polymer, respectively, and WP, EA, and FD stand for working point, extracting agent, and feed. The situation is illustrated in Fig. 6.5: Components in the feed and in the extracting agent must form a miscibility gap at the operational temperature. Compositions of feed and extracting agent must lie outside the miscibility gap in such a manner that their connection, called the working line, intersects the miscibility gap. The position of the working point on the working line inside the miscibility gap depends upon the flow rates of feed and extracting agent. The system disintegrates, according to the working point and relevant tie-line, in the sol and gel which leave the column. Phase triangles as in Fig. 6.5 can be determined by cloud-point measurements (see Sect. 6.6.2.3).

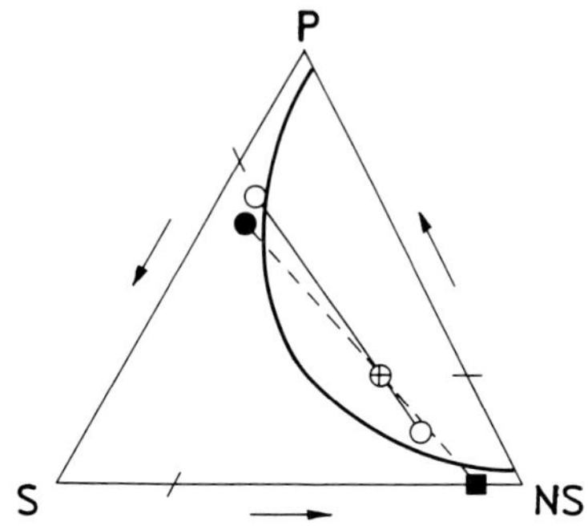


**Fig. 6.5.** Schematic representation of the continuous polymer fractionation in a phase triangle with solvent ($S$), nonsolvent ($NS$), and polymer ($P$) for $p$, $T$ = const.; *bold line* – cloud-point curve, ● – feed, ■ – extracting agent, ⊕ – possible working point, – – – working line, ○ – sol and gel, —— tie-line

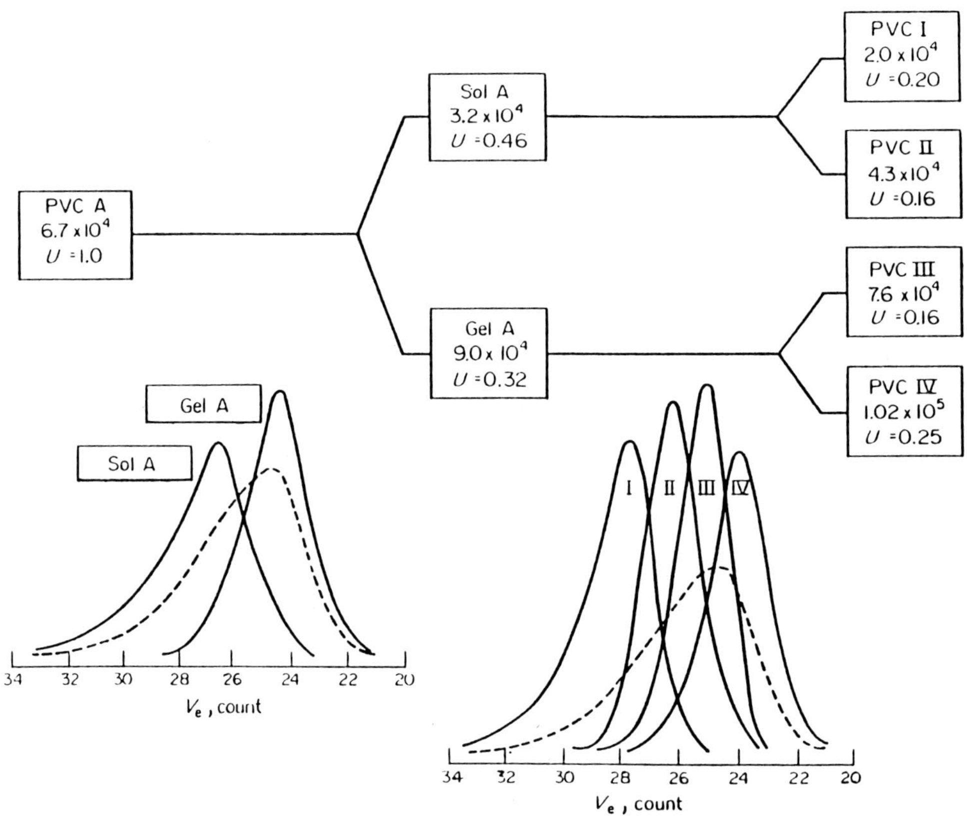


**Fig. 6.6.** Flow chart of a preparative continuous polymer fractionation of PVC. The gel produced in one fractionation step could be used without further treatment as feed in the successive extraction whereas the corresponding sol has to be precipitated and redissolved for that purpose. Weight-average MWs are indicated; *curves* are SEC results for initial polymer (*dashed*) and fractions. (Reprinted from Ref. [1], Comprehensive Polymer Science, ed. by Allen, vol. 1 (eds. Booth, Price), p. 293, Copyright 1989, with kind permission of Pergamon Press Ltd., Headington Hill Hall, Oxford OX3 OBW, UK)

When the column operates under stationary conditions, the following relations hold good:

$$\dot{V}^{FD} + \dot{V}^{EA} = \dot{V}' + \dot{V}'' \tag{6.3}$$

$$\dot{V}_i^{FD} + \dot{V}_i^{EA} = \dot{V}_i' + \dot{V}_i'' \tag{6.4}$$

$$m_i^{FD} + m_i^{EA} = m_i' + m_i'' \tag{6.5}$$

$$\bar{M}_w^{FD} \cdot m_P^{FD} = \bar{M}'_w \cdot m'_p + \bar{M}''_w \cdot m''_P \tag{6.6}$$

CPF as chromatographic counter-current extraction is a powerful tool for large-scale fractionations. It has been successfully used for polystyrene [1], poly(vinyl chloride) [13], poly(isobutylene) [15], phenol-formaldehyde resin [1], poly(ethylene) [16], poly(carbonate) [18], poly(vinyl methyl ether) [19], poly(dimethyl siloxane) [19], and poly(vinyl pyrrolidone) [19] up to the kilogram scale. The non-uniformities of the final fractions were about 0.2. Figure 6.6 gives an example for the fractionation of a 400 g sample of poly(vinyl chloride) with THF and water. Three runs were necessary to produce four fractions.

Instead of solvent/nonsolvent systems, temperature change by entering the column can be used to produce the coacervate from feed [16]. Likewise, pressure differences can be operating generally.

#### 6.6.2.2 Equipment and Materials

A crucial part of CPF is the fractionation column which will be discussed more detailed below. Additional equipment includes:

- thermostatted storage tanks of suitable volume for feed and extracting agent
- adjustable pumps for transport of feed and extracting agent through the column. Owing to the viscosity of the feed, an injection pump (one-plunger pump with continuous transport) of suitable volume is advantageous for the feed. Calibration of all pumps is recommended.
- flasks for collection of sol and gel leaving the column, and
- rotational-type evaporators for evaporation of solvent and non-solvent in sol and gel.

Devices for preparatory investigations (usually standard glassware, see next section) will not be discussed here.

**Types of Column**

Let us come back to the fractionation column. Three types have been used so far: pulsed sieve-bottom column [13(a), 15], packed column [17, 18], and mixer-settler apparatus [17]. Figure 6.7 shows schematically sieve-bottom and packed columns.

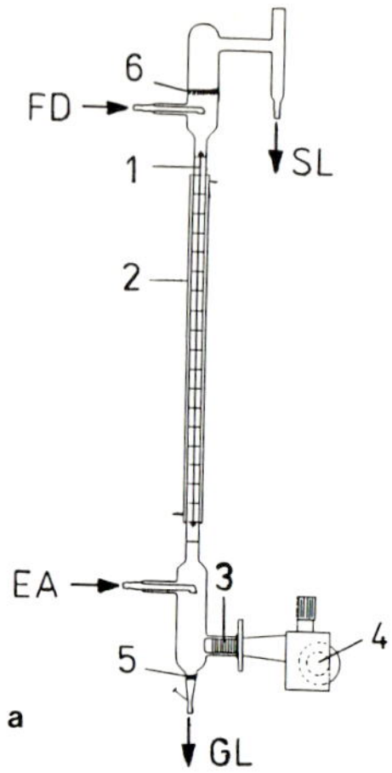


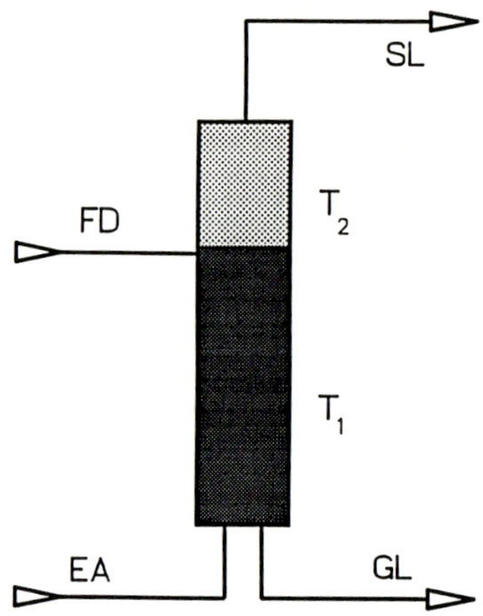


**Fig. 6.7a, b.** Columns for CPF (schematical). **a** Pulsed sieve-bottom column; *1* – sieve-bottom insertion, *2* – jacket for thermostatting, *3* – bellows for pulsation, *4* – pump for pulsation, *5, 6* – phase boundaries gel/mixed phase and sol/mixed phase. **b** Scheme of a packed column with lower extraction part ($T_1$) and upper "back-flow" part ($T_2$); usually holds $T_1 > T_2$. *FD* – feed, *EA* – extracting agent, *SL* – sol, *GL* – gel (Refs. [13(a), 17] with permission of Hüthig & Wepf Verlag, Basel)

**Sieve-bottom columns.** Thermostatted columns of 100 cm length and 2.5 cm [15] or 4 cm [13(a)] diameter, coupled with a pulsation device, were used. Effective extraction volumes are 500 and 1200 $cm^3$, respectively.

**Packed columns.** Packed columns filled with glass beads of 4 to 8 mm diameter result in higher fractionation efficiency than with sieve-bottom columns [17, 18]. The position of the feed inlet at about three quarters or two thirds of the total length is essential for the complete separation of sol and feed just as the temperature difference below and above of the feed inlet. The parts of the column above and below the feed inlet can be set to different temperatures (usually $T_1 > T_2$) [17, 18] or may have the same temperature [19]. In the latter case, a condenser with a lower temperature through which the sol flows must be mounted above the column. A condenser is also recommended when $T_1 > T_2$ in order to obtain sharp fractions. The usual dimensions of packed columns are: length 200 cm, diameter 3 cm [18]. Shorter columns are possible for the fractionation of smaller amounts of polymer [19] (see below).

Independent of the column type, some constructive aspects should be noted: The inlet tubes of the feed and extracting agent should be thermostatted between pump and column when the operating temperature and the ambient temperature are not the same or when the ambient temperature varies considerably. The feed inlet must be opened below as can be seen in Fig. 6.7a. (In a packed column, this inlet can be directly integrated into the complete column or, alternatively, the thermostatted inlet is arranged between the separate parts of the column and connected by tight-fitting ground-glass joints.) The inlet of the extracting agent must be at a higher level than the gel outlet which is placed in the lowest part of the column.

**Materials**

The following guidelines can be given for materials required (except glass beads for packed columns (see above), materials for further processing of sol and gel, and materials for preparatory investigations). Table 6.4 gives a rough compilation of data derived from literature. These values are only informative; conditions for special fractionations must be tested. Note that the given range of each value summarizes experimental conditions used. For individual fractionation runs, individual sets of values are valid, i.e., $\dot{m}_P^{FD} = 0.4$ g/min is not inevitably connected with $\dot{V}^{FD} = 3.0$ ml/min, $\dot{V}^{EA} = 7.5$ ml/min, and $t_{stat} = 2$ h.

**Polymer.** The required mass of polymer follows from mass flux of polymer in the feed, from fractionation time under stationary conditions, and from feed

**Table 6.4.** Compilation of required amounts of polymer, feed, and extracting agent depending on the type of CPF column (see text)

| Type of column | Sieve-bottom column | | Packed column | |
|---|---|---|---|---|
| $l$ (cm) | 100 | 100 | 100 [19] | 200 [18] |
| $d$ (cm) | 2.5 [14, 15] | 4.0 [13(a)] | 3.0 | 3.0 |
| Effective volume (cm$^3$) | 500 | 1200 | 138[f] | 276[f] |
| $\dot{m}_P^{FD}$ (g/min) | 0.4–0.7 | 0.4–0.5 | 0.02–0.05 | 0.07–0.2 |
| $m_P^a$ (g/h) | 24–42 | 24–30 | 1.2–3 | 4–12 |
| $m_P^b$ (g) | 90–375 | 90–450 | 20–30 | 30–110 |
| $\dot{V}^{FD}$ (ml/min) | 3.0–5.0 | 3.0 | 0.1–0.3 | 0.25–1.4 |
| $V^{FDc}$ (ml/h) | 180–300 | 180 | 6–18 | 15–85 |
| $V^{FDd}$ (ml) | 600–2500 | 600–3000 | 150–200 | 200–750 |
| $\dot{V}^{EA}$ (ml/min) | 7.5–40[g] | 10–60 | 1–2 | 3.5–5.5 |
| $V^{EAc}$ (ml/h) | 450–2400 | 600–3600 | 60–120 | 210–330 |
| $\dot{V}^{EAd}$ (ml) | 2500–4400 | 9000–11400 | 1200–1350 | 2000–2600 |
| $t_{stat}^e$ (h) | 2–8 | 4–15 | 10–20 | 7–12 |

[a] Mass of polymer possible to fractionate within one hour after stationary conditions were reached.
[b] Mass of polymer necessary to reach stationary conditions ($w_P^{FD} = 0.15$ and $\rho^{FD} = 1.0$ g/cm$^3$ are assumed).
[c] Feed or extracting-agent volume related to one hour (stationary conditions).
[d] Feed or extracting-agent volume necessary to reach stationary conditions.
[e] Time approximately necessary to reach stationary conditions.
[f] Volume below the feed inlet is assumed here as 75% of the total interstitial volume.
[g] Frequently used value is about 25 ml/min.

volume necessary to adjust stationary conditions (see below). These parameters are influenced by type and dimensions of the column as can be seen in Table 6.4: Sieve-bottom columns work with higher polymer fluxes than packed columns with comparable dimensions and therefore, the polymer mass fractionated per hour is quite different. The polymer required for reaching stationary conditions is smaller using packed columns.

**Feed.** The total feed volume can also be divided in the nearly fixed part (related to a special column and certain working conditions) necessary to reach stationary conditions, and the variable part depending on flow rate $\dot{V}^{FD}$ and fractionation time. Concrete $\dot{V}^{FD}$ values also depend on the ratio $\dot{V}^{EA}/\dot{V}^{FD}$. Packed columns usually need smaller flow rates. Volumes of solvent and nonsolvent in the feed can be calculated according to the feed composition. The feed volume necessary to adjust stationary conditions is connected with the *effective* volume of the column (see Table 6.4) by the ratio $\dot{V}^{EA}/\dot{V}^{FD}$. The greater this ratio the smaller is the required volume of the feed (and, depending on polymer concentration in the feed, the smaller the amount of polymer needed for reaching stationary conditions, see above). The effective volume in the sieve-bottom column is almost the whole extraction part (between the points of mixing and separation of the components inside the column) whereas this volume in a packed column is represented only by the interstitial volume (26% of the total volume). Therefore, the effective volumes of the columns are quite different. Stationary conditions can be reached at the latest after a tenfold exchange of the effective volume (note the control of stationary conditions by SEC, described in Sect. 6.6.2.3). Required feed volumes and polymer masses in Table 6.4 were estimated with this assumption in consideration of different ratios $\dot{V}^{EA}/\dot{V}^{FD}$ and therefore represent the highest values. However, it should be noted that gel and sol fractions often already have sufficient sharpness in the nonstationary running so that they can be used for further investigations. Besides, the "nonstationary" gel can be led back as feed [19].

**Extracting agent.** The aspects outlined for the feed hold just as well for the extracting agent. Usually, flow rates and required volumes of the extracting agent are higher than that of the feed. Leading back of "nonstationary" sol as extracting agent is under discussion [19].

Finally, data in Table 6.4 give a rough estimation of the maximum period of time for reaching stationary conditions. Packed columns need longer periods.

#### 6.6.2.3 Preparatory Investigations

Extensive investigations are necessary to guarantee a successful CPF run. Let us discuss this investigations in their logical succession.

**Choice of the Solvent/ Nonsolvent System**

As usual, thermodynamically poor solvents and weak precipitants should be chosen. The dissolution power of the solvent must, however, prevent association of the polymer. Besides, the following properties of solvent and nonsolvent are advantageous: differences in densities of solvent, nonsolvent, and polymer so that the flows through the column (FD → GL, EA → SL, see Fig. 6.7) are aided, i.e., for normal flow $D^{FD} > D^{EA}$ should hold; low viscosity and surface tension facilitate a good intermixing of feed and extracting agent.

**Choice of Column**

Type and dimensions of the column influence the required amounts of polymer, solvent, and nonsolvent, likewise the time necessary for reaching stationary conditions and finally the overall fractionation time (cf. Table 6.4). The smaller the mass of polymer the smaller the column should be. Packed columns usually have a higher fractionation efficiency but they work slower than sieve-bottom columns of similar dimensions. For the latter, suitable frequency and amplitude of pulsation must be tested (mostly used frequencies were 3.5 to 4.5 $s^{-1}$ [13(a), 14, 15]).

**Phase Triangle and Tie-lines**

Knowledge of the phase diagram and the corresponding tie-lines at fractionation temperature (see Fig. 6.5) are necessary to decide the experimental conditions. First, the miscibility gap must be determined by cloud-point measurements (turbidimetric titration, see Sect. 5.4.2) on polymer solutions of varying concentration. Tie-lines connecting the conjugated phases (cf. Fig. 3.2) can be found by demixing of ternary systems with overall composition inside the miscibility gap and analysis of the originated phases. Cloud-point curve and conjugated phases cannot coincide exactly because the former is determined using the unfractionated sample whereas the points of sol and gel phase are caused by the MWD of these fractions. Now, several working parameters must be determined.

**End of Fractionation**

One has to decide the number of fractions and the MW where the distribution cutoff will be. The planes left and right from the cutoff in the MWD direct the desired mass fractions of the polymer contained in sol and gel, respectively, and thereby the $\dot{G}$ value necessary in this CPF run (cf. Eq. (6.1)).

**Determination of Parameters**

**Determination of working point and the corresponding *G* value.**
In order to find the working point and the corresponding $G$ value ($G$ instead of $\dot{G}$ represents static experiments), demixing of ternary mixtures (polymer, solvent, nonsolvent) is carried out and the polymer mass in the phases is determined. The following factors influence the relationship of working point and $G$ value (cf. Fig. 6.8): $G$ is zero when the miscibility gap is entered from high polymer concentrations and $G$ is infinite when the two-phase region is left in the direction of the pure extracting agent. Consequently, $\dot{G}$ ascends with

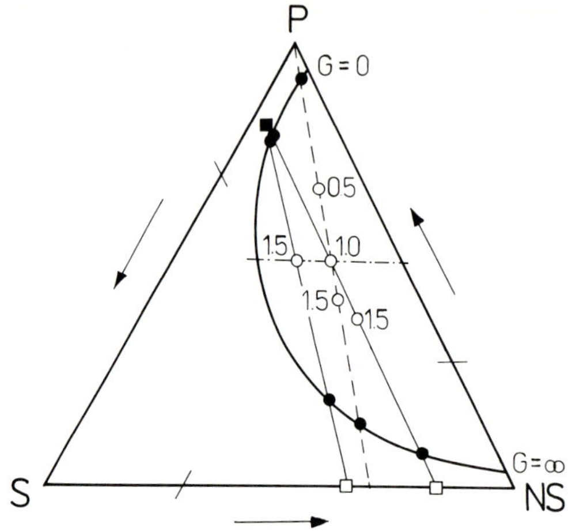


**Fig. 6.8.** Relation of working point and $G$ value (schematical). —— miscibility gap, —— working lines, – – – line with constant S/NS ratio, –·–·– line with constant overall polymer concentration, ■ – feed, □ – extracting agents, ● – points with $G = 0$ or $\infty$, ○ – working points. *Numbers represent fictitious G values*

decreasing overall polymer concentration (which can be the result of both the decreasing polymer fraction in the feed and the increase of the flow ratio $\dot{V}^{EA}/\dot{V}^{FD}$). On a line of constant overall polymer concentration, $G$ diminishes when the nonsolvent fraction in the system increases. The fractionation efficiency for working points having identical $G$ values increases on approaching the coexistence curve. The fractionation efficiency to be expected in CPF can be estimated by SEC (or other MW determinations) of the polymer in the conjugated phases.

**Choice of feed composition, working line, and extracting agent.**
The feed must lie outside the miscibility gap at relatively high polymer concentration (often 15%). However, this concentration point should be near to the coexistence curve. A distance of 1% from the miscibility gap, applied to the solvent scale, is favourable. Smaller distances may lead to demixing of the feed on varying room temperature. Working point and feed composition fix the direction of the working line and the composition of the extracting agent. The feed should be chosen so that the working line and tie-line are nearly in the same direction.

**Calculation of flow rates of feed and extracting agent.**
With the known values for $w_i^{WP}$, $w_i^{EA}$, and $w_i^{FD}$ (i = solvent, nonsolvent, polymer), the ratio of flow rates $\dot{V}^{EA}/\dot{V}^{FD}$ can be calculated using Eq. (6.2) or can be estimated graphically via the distances of the concentration points of the extracting agent, working point, and feed in Fig. 6.5. The absolute values depend on system and column parameters.

**Test and calibration of the pumps.**
See p. 87.

### 6.6.2.4 Fractionation Procedure

**Materials**

**Preparation of the feed.**
The polymer is dissolved in the pure solvent component according to concentration, composition, and required volume. The nonsolvent is added after complete dissolution. Repeated monitoring of the composition (especially for low-boiling liquids) by weighing, refractometry, or densitometry is recommended. For very high-MW polymers, lowering of the concentration in the feed is recommended [14].

**Preparation of the extracting agent.**
Solvent and nonsolvent are mixed according to phase triangle. The composition should also be checked as mentioned for the feed.

**Equipment Setting up**

The column (empty sieve-bottom column or packed column filled with glass beads) is connected with pumps and storage tanks filled with feed and extracting agent, respectively. The whole apparatus is thermostatted, if necessary. The condenser above the packed column should be set to a temperature a few Kelvin (often 10 K) lower than the column. The lower the temperature of the condenser the lower is the $\dot{G}$ value. The best variant must be tested.

**Operations**

**Start of fractionation.**
First, extracting agent is pumped into the column up to the feed inlet. Then, calculated flows are adjusted and the flows of feed and extracting agent are started. The pulsation of a sieve-bottom column must be switched on.

**Operating the column.**
Care must be taken that the flows are uniform and constant and the development of the phase boundaries on top and foot of the column should be monitored. Damming up of the gel may arise when the flows are too fast (or the used glass beads too small). Very high-MW polymer samples also require a lowering of the rate of flow [14].

**Stationary conditions.**
The column conditions are stationary when the SEC results of the polymer fractions in sol and gel phases remain constant.

**Collection and Processing**

**Sol and gel fractions.**
The gel fraction must be collected at such a rate that the boundary of the gel is always below the inlet of the extracting agent (cf. p. 89. Solvent and nonsolvent of sol and gel are evaporated by use of rotational-type evaporators and the polymer is dried in a vacuum. If necessary (SEC control!), collected portions of the same phase (fraction) can be re-dissolved and re-precipitated together in order to obtain greater homogeneity.

Optimal results in CPF can be expected usually only after several test runs.

## 6.7 Examples of Extraction Fractionations

Information given in Sects. 4.1, 4.2, and 4.3.1 should be noted.

### 6.7.1 Film-Extraction Fractionation of Poly(vinyl Acetate) with Methyl Acetate/Petrol Ether[3]

The given specifications for volumes of vessels and solvents or nonsolvents, and for support surfaces can be converted linearly for polymer masses deviating from the mentioned ones. See also Sect. 6.3.

**Materials**

**Polymer.** Poly(vinyl acetate), polymer mass depends on coating procedure, see notes on Fractionation in this section.

**Solvent.** Methyl acetate, volume for coating depends on procedure, see notes on Fractionation in this section; 0.5–1 l for extraction steps

**Nonsolvent.** Petrol ether, 0.5–1 l for extraction steps, 100 ml/100 mg polymer fraction to be precipitated

**Support.** 1000–1500 cm$^2$ aluminum foil or cotton textile.

**Equipment**

- Conical flask (50 or 200 ml, see notes on Fractionation in this section) for dissolution of the polymer
- Shallow glass basin for coating the support
- Ventilator and exhauster
- Extraction vessel as shown in Fig. 6.1 (250 ml, advantageously equipped with thermo-jacket and connected with thermostat and recirculating pump), or conical flask (about 250 ml) and thermostat
- Mechanical shaker
- Flask (250 ml) with ground-glass joint to take up the extracted sol phase (required for each fraction)
- Syringes
- Rotational-type evaporator
- Glass beaker (about 250 ml, for 100 mg polymer fraction to be precipitated)
- Stirrer
- Small dropping-funnel
- Shallow glass dishes
- Vacuum oven
- Stands, clamps, sockets, etc.

[3] For this example, data given by Fuchs [4] were used.

**Time Required**

- Dissolution of the polymer (2–3 h) overnight
- Coating (depending on procedure) 4–5 h
- Drying of the coated material overnight
- Extraction and isolation (each fraction) 0.5 h

Time necessary for processing of fractions is not included.

**Preparatory Investigations**

- Check of complete solubility of the polymer in the solvent
- Determination of the limits of extraction: cloud-point by addition of nonsolvent to the polymer solution (*c* about 1 g/l); solvent/nonsolvent composition which leads to initial solubility of the polymer by checking the test mixtures with excess nonsolvent
- Choice of the extraction steps from the limits related to the number of fractions

**Fractionation**

Fractionation temperature: 25 °C

- Dissolution of PVAC with shaking inside a conical flask. Polymer mass and solvent volume depend on coating procedure. Possible variants: 10 g PVAC in 100 ml or 1 g PVAC in 10 ml
- Coating the support can be performed in three ways:

- Dipping of aluminum foil with an *overall* surface area of about 1000–1500 $cm^2$ (in convenient pieces) into the solution (10 g, 100 ml); drying in air using ventilator and exhauster; weighing about 1 g PVAC should be on the surface; the dipping must be repeated, if necessary. (Note that, usually, the polymer film still contains a few percent solvent even after air drying.)
- Pouring the PVAC solution (1 g, 10 ml) onto the aluminum foil which is bent up around the solution; drying in air, see first method. Here, one needs smaller and specific amounts of polymer and solvent but, the aluminum foil must be twice as large as in the first method.
- Absorption of the PVAC solution (1 g, 10 ml) by cotton textile – the solution can be taken up almost completely; drying in air, see first method. Depending of the texture of the cotton material, careful dilution of the solution before absorption may be advantageous.

- Cutting the dried support in convenient pieces which can be introduced into the fractionation vessel (or conical flask) in coiled or crumpled form (protection of the coating against being rubbed off!)
- Extraction with slight shaking with 100 ml of the solvent/nonsolvent mixture made up according to the preparative investigations for each fraction; extraction time is about 10 minutes. The last fraction should be extracted with pure solvent.
- Isolation of the sol fraction by draining or by careful decanting into a suitable vessel. Care must be taken so that there are no dislodged polymer

particles! If necessary, filtration and return of the filter to the next extraction step may be tried. The coated foil should be washed with a small volume of the used extraction mixture before the next extraction step.
- Evaporation (concentration) of the sol phases (< 10 ml) using the rotational-type evaporator
- Precipitation into 100 ml ice-cooled petroleum ether. PVAC yields a soft precipitate which can only be filtered with difficulty. Therefore, repeated extraction of the precipitate with petroleum ether is recommended.
- Drying in shallow glass dishes in a vacuum (30–40 °C).

**Evaluation**

- Determination of mass and MW or related values ([$\eta$], $\varphi^*$, $V_e$ from SEC) for each fraction
- In an ideal case, all fractions of a fractionation should have the same mass. A plot of the cumulative mass of fractions vs. solvent/nonsolvent composition in each extraction step (*extraction curve*) makes it possible to optimize a repeat of the fractionation.
- Comparison of directly measured and calculated averages of MW or [$\eta$] for the whole sample according to Eqs. (2.2) to (2.4) and (4.1)
- Comparison of gradation and sharpness of fractions and starting polymer by SEC or turbidimetric titration

### 6.7.2 Column-Extraction Fractionation of Poly(styrene) with Butanone/Methanol

This example is suggested for different scales of columns and polymer masses. Table 6.5 summarizes some operating variables in dependence on column dimensions. Other values can be interpolated. See also Sect. 6.4.

**Materials**

**Polymer.** Poly(styrene), $\bar{M}_n$ about $10^5$ g/mol, $U = 2$–3; masses see Table 6.5.

**Solvent.** Butanone (analytical grade), 50 ml/1g polymer for coating the support. Volume of extraction steps depends on composition and volume of the extracting agents of the individual steps (see Table 6.5).

**Nonsolvent.** Methanol (analytical grade), interstitial volume (see Table 6.5) in order to fill the column before the fractionation starts. For the individual extraction steps, the same aspects hold as for butanone (see Table 6.5).

**Support.** Glass beads or sea sand ($d = 0.2$–0.5 mm), about 50 cm$^3$/1 g polymer. Calculations were performed for glass beads with diameter of 0.4 mm and support load of 22.2 mg/cm$^3$ (see Table 6.5).

Solvents and nonsolvents for precipitation, re-dissolution, and re-precipitation of the fractions must be related to their masses. Approximate

**Table 6.5.** Operating variables of column extractions (Expl. 6.7.2)

| Column dimensions: $l$ (cm) | 20 | 50 | 100 | 50 | 100 |
|---|---|---|---|---|---|
| $d_i$ (cm) | 2.5 | 2.5 | 2.5 | 7.0 | 7.0 |
| Support (bulk volume) ($cm^3$) | 100 | 245 | 490 | 1900 | 3800 |
| Polymer[a] (g) | 2.2 | 5.4 | 10.8 | 42 | 84 |
| Interstitial volume (appr.) (ml) | 26 | 64 | 128 | 495 | 990 |
| Flow rate (ml/min) | 2 | 2 | 2 | 20 | 20 |
| Eluate volume per step[b] (ml) | 60 | 150 | 300 | 1000 | 2000 |
| Time per step[b] (h) | 0.75 | 1.5 | 2.75 | 1.1 | 2.0 |

[a] Values are related to glass beads of diameter 0.4 mm and a load (polymer/bulk volume) of 22.2 mg/cm$^3$ (=12.1 mg/g = 0.2 mg/cm$^2$).
[b] After flow of the interstitial volume of eluent (extracting agent) into the column, elution is broken off for 15 minutes and then continued with the rest of extracting agent (see notes on fractionation in this section).

quantities: $\leqslant$100 ml solvent/1 g polymer (re-dissolution), $\geqslant$1 l nonsolvent/ 1 g polymer (precipitation or re-precipitation).

**Equipment**

- Glass beaker (250 ml) for coating the support. Greater volumes of support should be coated in several portions with an identical polymer/ solvent/support ratio.
- Glass rod
- Ventilator and exhauster
- Mortar
- Column as shown in Fig. 6.2 furnished with thermo-jacket, recirculating thermostat, degasser, storage vessel (1 l), and valve. Use in normal-flow direction. Column dimensions are given in Table 6.5. When the column tube is only partly filled with the coated support, the unfilled part can be used as degasser.
- Collecting vessels for fractions. Number and volumes depend on number and volume of sol fractions. A fraction collector may be useful.
- Rotational-type evaporator for concentration of the sol phases
- Glass beaker for precipitation or re-precipitation of fractions (>1 l/1 g polymer)
- Stirrer
- Dropping funnel
- Device for filtration under vacuum or pressure (filter from sintered glass or paper)
- Shallow glass dishes
- Vacuum oven
- Stands, clamps, sockets, etc.

**Time Required**

- Dissolution of polymer 2–3 h
- Coating the support (one portion) 2–3 h
- Filling and assembly of column 1 h
- Thermostatting 1 h
- Extraction time per step (cf. Table 6.5) 0.75–3 h
- Purge of the packed column 0.5–2 h

Times necessary for cleaning the support and processing of the fractions are not included.

**Preparatory Investigations**

See Sect. 6.7.1. The extraction steps (solvent/nonsolvent composition) should not be of uniform size: they should decrease with increasing MW.

**Fractionation**

Fractionation temperature: 25 °C

- Cleaning the support by washing with concentrated hydrochloric acid (exhauster!), distilled water (repeatedly), and butanone
- Dissolution of the polymer in butanone, concentration about 2–5%
- Addition of the corresponding amount of support (see Table 6.5) to the polymer solution, gentle heating with stirring by hand and evaporation of the solvent (ventilator and exhauster)
- Careful crushing of clumps; quantitative transfer of the coated support from the beaker
- Filling the column (after the outlet is closed) with degassed nonsolvent according to the interstitial volume (see Table 6.5) and afterwards with the coated support. The support must be always covered with liquid.
- Setting up the column according to Fig. 6.2 (normal flow)
- Filling the degasser with nonsolvent and the storage vessel with the first solvent/nonsolvent mixture (closed stopcock)
- Thermostatting of column and degasser
- Start of the extraction by connecting the storage vessel with the column and regulation of the flow rate (see Table 6.5) with the valve
- Passing of about half the eluate volume given in Table 6.5 through the column (the interstitial volume of nonsolvent is now replaced by the first extraction mixture), suspension of elution for 15 minutes
- Flow of the rest of the first extraction mixture into the column (till the storage vessel is empty), collection of the volume flowing out of the column
- Filling the storage vessel with the next extraction mixture, replacement of the first extraction mixture by the next (interstitial volume); the corresponding eluate volume belongs to the first fraction
- No elution for 15 minutes; change of the vessel collecting the sol phase
- Introduction of the rest of this extraction mixture into the column in the same manner as for the elution of the first fraction; collection of the second fraction

- Repeat of the procedure for all fractions. The collection of a new fraction always starts with the end of the flow break. (Elution without a break consumes more solvent and nonsolvent per step in order to extract a fraction almost completely.)
- Elution of the last fraction with a plenty of pure solvent (purge of support and column)
- Evaporation of sol phases using a rotational-type evaporator to concentrate the fractions for precipitation (about 1%) or to isolate the dry fractions
- Precipitation of the concentrated sol phases by dropping into a tenfold volume of methanol. Re-dissolution and re-precipitation are possible
- Drying of the fractions under vacuum (about 50 °C).

**Evaluation**

See Sect. 6.7.1

### 6.7.3 Coacervate-Extraction Fractionation of Poly(ethylene) in Decahydronaphthalene/Dibutyl Phthalate by Increasing Temperature

This example should be compared with Sect. 5.5.4.

**Materials**

**Polymer.** Poly(ethylene) (10 g) of low (LDPE) or high density (HDPE)

**Solvents.** Decahydronaphthalene (DHN), 400 ml for the starting polymer, about 200–300 ml per extraction step; Xylene, 100 ml/1 g polymer fraction for re-dissolution of sol phase

**Nonsolvents.** Dibutyl phthalate (DBP), 600 ml for the starting polymer, about 300–400 ml per extraction step
Methanol (analytical grade), 1 l/1 g polymer fraction for precipitation
Methanol or acetone for Soxhlet extraction

**Stabilizer.** (e.g., phenyl-$\beta$-naphthyl amine), about 0.1 g

**Equipment**

- Thermostat (oil bath) equipped with glass window, stirrer, and temperature-control unit (about 20–160 °C) or, alternatively, extraction vessel with thermo-jacket and recirculating pump
- Three-neck vessel (2 l) similar to that shown in Fig. 5.1a–d
- Tight-fitting stirrer
- Thermometer or thermoelectric couple (about 0–200 °C)
- Dropping funnel
- Inlet tube for nitrogen
- Capillary tube with ground-glass joint (in place of a stopper)
- Ground-glass stoppers

- Pipettes or syringes
- Flask (1 l) to take up sol phases
- Rotational-type evaporator
- Conical flask (about 200 ml, for 1 g polymer fraction) with ground-glass joint and stopper to take up the re-dissolved sol fraction
- Glass beaker (>1 l, for 1 g polymer fraction to be precipitated)
- Stirrer and dropping funnel
- Filtration device for vacuum or pressure
- Soxhlet extractor
- Vacuum oven
- Stands, clamps, sockets, etc.

**Time required**

| | |
|---|---|
| – Thermostatting | 1 h |
| – Dissolution of starting polymer | 2–3 h |
| – Cooling (coacervate) | 3–5 h |
| – Phase separation | (if necessary) overnight |
| – Phase isolation | 1 h |
| – Addition of extracting agent and temperature rising (dependent on temperature difference) | 1–3 h |
| – Stirring at constant temperature | 0.5 h |

Except for steps one to three, all periods are valid for each fraction. The time necessary for processing of fractions is not included.

**Preparatory investigations**

- Check of complete solubility at elevated temperature and determination of the cloud-point temperature by slow cooling
- Determination of the temperature of first solubility by stepwise heating (e.g., 5 K) and checking the clear supernatant with excess methanol (it can be that cooling is sufficient to give turbidity)
- The interval which has been determined should be divided into temperature steps of decreasing size (e.g., 30, 20, 13, 8, . . . K) depending on the number of fractions.

**Fractionation**

- Setting up of the fractionation device: extraction vessel (inside the thermostat or with thermo-jacket) equipped with a tight-fitting stirrer and nitrogen inlet/outlet
- Heating of 400 ml DHN to about 150–160 °C, addition of the stabilizer (with stirring)
- Dissolution of 10 g PE in the hot solvent with stirring and flushing nitrogen. The required dissolution temperature may be a few Kelvin higher for HDPE than for LDPE.
- Slow addition of 600 ml hot DBP, with stirring, to the solution; the mixture must remain clear.

- Replacement of the nitrogen inlet and outlet with a thermometer, or thermoelectric couple and a capillary tube; slow cooling with stirring near to the temperature of first solubility (cf. Preparatory investigations)
- Rise in temperature to about 50–60 °C. The following temperature intervals should be reduced successively.
- Stirring for about 30 min at constant temperature
- Separation of the phases at constant temperature. Owing to the density of PE, the gel phase (coacervate) is here the upper phase.
- Isolation of the sol phase can be performed by extraction with a syringe or by draining the lower phase (under flushing nitrogen). The polymer can precipitate during the isolation when temperature differences exist.
- The isolated sol phase is substituted by a similar volume of the DHN/DBP (4:6) mixture. Then, the temperature is raised (under stirring) in order to obtain the next fraction which can be isolated after separation.
- Concentration or drying of the sol fraction by vacuum evaporation using a rotational-type evaporator
- Re-dissolution of the concentrated or dry sol phase in the flask (and in the syringes!) with xylene at about 120 °C and dropping into a tenfold volume of methanol with stirring
- Soxhlet extraction of the filtered fraction with acetone or methanol, drying in a vacuum at about 50 °C.

The number of fractions and the temperature intervals in each fractionation step must be estimated in connection with the masses of the foregoing (dried) fractions.

**Evaluation**

See Sect. 6.7.1. The extraction curve can be obtained by plotting the cumulative mass of fractions vs. extraction temperatures. The volume of the extracting agent used also influences the polymer mass in each fraction. MWDs of polyolefines often correspond to the logarithmic normal distribution (cf. Sect. 4.3.3 and Table A 7). The result (for the same polymer) can be compared with Sect. 5.5.4.

### 6.7.4 CPF of Poly(isobutylene) with Toluene/Methanol[4]

**Materials**

**Polymer.** Poly(isobutylene), $\bar{M}_w$ ca. $10^5$ g/mol, polymer mass (about 50 g) depends on the time of fractionation (cf. Table 6.4)

**Solvent.** Toluene (technical grade), the volume required (about 3 l) depends on the working parameters and fractionation time (cf. Table 6.4).

[4] This example was prepared using information kindly given by Prof. B.A. Wolf, K. Schultes, and R. Mertsch. This help is gratefully acknowledged.

**Nonsolvent.** Methanol (technical grade), the volume required (about 5 l) depends on the working parameters and fractionation time (cf. Table 6.4).

**Equipment**

For preparatory and control investigations are necessary:

- Device for cloud-point or turbidimetric-titration measurements
- Several separating funnels for phase investigations
- Devices for phase analyses (including determination of the polymer masses in the phases)
- SEC apparatus to monitor the efficiency of the working parameters and the stationary conditions

**Equipment for fractionation**[5]

- Columns ca. 65 cm (1) and ca. 20 cm (2) long with a diameter of about 3 cm. These columns are either connected via ground-glass joints (sealed with teflon foil) or directly fused with the feed inlet (between the columns), extracting-agent inlet and gel outlet (on the foot of column 1, extracting-agent inlet some higher than gel outlet), and condenser.
- Condenser about 20 cm long, the top connected to the sol outlet and the bottom to the top of column 2. The temperature should be set 10 K below the operating temperature of the columns.
- Glass beads with diameters of about 6 mm (column 1) and about 4 mm (column 2)
- Pumps for feed and extracting agent with an adjustable flow range of 0.1 to about 2 ml/min
- Storage tanks with a suitable volume for the feed and the extracting agent, connected with a pump and feed or extracting-agent inlet, respectively
- Flasks with ground-glass joints to collect sol and gel leaving the column
- Rotational-type evaporators.

**Time required**

- Dissolution of polymer (feed preparation) — overnight
- Reaching stationary conditions ($\dot{V}^{FD} = 0.2$ ml/min, $\dot{V}^{EA} = 2.0$ ml/min; maximum time is given) — about 10 h
- Fractionation (after reaching stationary conditions) — variable

Times for preparatory investigations, emptying and cleaning of the column, and processing of fractions are not included.

**Preparatory investigations**

- Determination of the miscibility gap by addition of methanol to toluene solutions of PIB of different concentrations (about 1 to 20% PIB)
- Determination of two or three tie-lines by demixing experiments and analysis of the conjugated phases

[5] Thermostatting of columns, storage tanks, and inlets of feed and extracting agent can be omitted when the CPF is performed under almost constant room temperature.

- Choice of feed composition, direction of the working line (extracting-agent composition), and working point. In order to achieve a relatively short time necessary for stationary conditions, one should choose the feed composition ($w_P^{FD} = 0.15$, $w_S^{FD} = 0.01$ higher than the corresponding value of the coexistence curve) and a few extracting agents of different $w_S^{EA}/w_{NS}^{EA}$ ratios to obtain different working lines. Equation (6.2) yields the working point of each working line using $\dot{V}^{FD} = 0.2$ ml/min and $\dot{V}^{EA} = 2.0$ ml/min for calculation purposes. $G$ values for each working point can be determined by demixing experiments as performed for tie-lines. A suitable working point can be chosen according to Fig. 6.8 by slight variation of the compositions of feed or extracting agent.
- SEC analyses of the unfractionated polymer and of the polymer fractions of the conjugated phases are recommended.

**Fractionation**

Fractionation temperature: room temperature.
Calibration of the pumps and assembly of the empty column (glass beads inside) with inlets and outlets have already been completed.

- After dissolution of PIB sample overnight in toluene, methanol addition to the chosen concentration next morning (check composition!)
- Preparation of extracting agent inside the storage tank (check composition!)
- Thermostatting of column, inlets of feed and extracting agent, and storage tanks (can be omitted when the room temperature is almost constant)
- Setting the temperature of the condenser 10 K below the fractionation temperature
- Filling column 1 with extracting agent, stop just below feed inlet
- Start of the feed (0.2 ml/min) and extracting-agent flows (2.0 ml/min)
- Monitor sol and gel phase leaving the column by SEC in order to find out when stationary conditions are reached
- Fractionation of the polymer under stationary conditions
- Simultaneous collection of sol and gel
- Evaporation of solvent and nonsolvent of sol and gel
- Cessation of the flows of feed and extracting agent when the CPF run is complete.

**Evaluation**

- Determination of mass, MW, and MWD of sol and gel fractions
- Deduction of improved working parameters.

## References

1. Koningsveld R, Kleintjens LA, Geerissen H, Schützeichel P, Wolf BA (1989) In: Booth C, Price C (eds) Comprehensive polymer science, vol. 1. Pergamon, Oxford, p 293

2. Kamide K (1977) In: Tung LH (ed) Fractionation of synthetic polymers. Marcel Dekker, New York, p 103
3. Elliot JH (1967) In: Cantow MJR (ed) Polymer fractionation. Academic, New York, p 67
4. Fuchs O (1950) Makromol Chem 5: 245; (1951) Makromol Chem 7: 259; (1956) Z Elektrochem 60: 229
5. Blair DE (1970) J Appl Polym Sci 14: 2469
6. Schröder E, Müller G, Arndt KF (1988) Polymer characterization. Hanser, Munich; (1989) Polymer characterization. Akademie-Verlag, Berlin
7. Englert A, Tompa H (1970) Polymer 11: 507
8. Theil I, Calugaru EM, Feldman D (1972) Faserforsch Textiltechn 23: 123
9. Huggins ML, Okamoto H (1967) In: Cantow MJR (ed) Polymer fractionation. Academic Press, New York, p 1
10. Koningsveld R (1970) Adv Polym Sci 7: 1
11. Holtrup W (1977) Makromol Chem 178: 2335
12. Porter SR, Johnson JF (1967) In: Cantow MJR (ed) Polymer fractionation. Academic Press, New York, p 95
13. Geerissen H, Roos J, Wolf BA (1985) Makromol Chem 186: (a) 735; (b) 753; (c) 769; (d) 777; (e) 787; (f) 801
14. Geerissen H, Roos J, Schützeichel P, Wolf BA (1987) J Appl Polym Sci 34: 271
15. Geerissen H, Schützeichel P, Wolf BA (1987) J Appl Polym Sci 34: 287
16. Geerissen H, Schützeichel P, Wolf BA (1990) Makromol Chem 191: 659
17. Wolf BA (1992) Makromol Chem, Makromol Symp 61: 244
18. Weinmann K, Wolf BA, Rätzsch MT, Tschersich L (1992) J Appl Polym Sci 45: 1265
19. Wolf BA (priv. comm.)

# 7 Gradient-Elution Fractionation

The technique of gradient-elution fractionation devised by Desreux et al. [1] has been developed to the most frequently and most variably used fractionation procedure [2–6]. Let us first consider some problems of the generation and utilization of gradients in fractionation procedures.

## 7.1 Types and Generation of Gradients

The use of gradients with continuously increasing solvent power offers two advantages over a stepwise fractionation: The fractionation takes place continuously which favours automation, and the mass balance of the fractions can be achieved very easily by subsequent combination of some neighbouring fractions which can be collected as relatively small portions. Continuously increasing the solvent power can be done by changing the solvent/nonsolvent ratio or the elution temperature. Gradients of this kind are described by the change of eluent composition or elution temperature dependent on the elution volume or the duration of elution.

### 7.1.1 Solvent/Nonsolvent Gradients

The course of a solvent/nonsolvent gradient (composition of the elution mixture vs. elution time or volume) can be linear [7], quadratic, concave, convex, S-shaped [8], logarithmic (exponential) and others. With polymers, logarithmic gradients are advantageous for separation by MW. Generation and use of gradients are described in Refs. [7–12]. Special mixing vessels of different shape or, nowadays, modern gradient devices as in HPLC can be used for the generation of different gradients. Let us restrict ourselves to the logarithmic gradient.

Its generation is easy. A *logarithmic gradient* originates when the volume which flows into the column is, in a mixing vessel equipped with a stirrer, replaced by an equal volume of the stronger solvent "B" from the storage vessel. Initially, the mixing vessel is filled with the starting mixture "A" (nonsolvent). Such a gradient is relatively steep at the beginning (when the lowest MWs are eluted) and becomes flatter with increasing solvent power.

This is advantageous for fractionations in the high-MW range. The logarithmic gradient can be described by the following equation:

$$\varphi_{m,t} = \varphi_{st} + (\varphi_{m,0} - \varphi_{st})\exp(-\dot{V}\cdot t/V_m) \qquad (7.1)$$

with $\varphi_{m,t}$ – volume fraction of solvent (or nonsolvent) in the mixing vessel at time $t$, $\varphi_{m,0}$ – the same at time 0 (mixed precipitant "A"), $\varphi_{st}$ – volume fraction of solvent (or nonsolvent) in the storage vessel (mixed solvent "B", $\varphi_{m,t}$ at $t=\infty$), $\dot{V}$ – flow rate of the eluent, $V_m$ – volume of the mixing vessel, $t$ – elution time.

Equation (7.1) shows that $\varphi_{m,t}$, i.e., the composition of the eluent at any time and thereby the course of the gradient, depends on all other quantities. Table 7.1 summarizes results of four variants of the parameter set (the latter are shown in the upper part). From the results in the lower part of Table 7.1 the following conclusions can be drawn:

- Increasing differences $\varphi_{m,0} - \varphi_{st}$ lead to steeper gradients (cf. variants A and B).
- Growing flow rate yields also a steeper gradient. A certain value of $\varphi_{m,t}$ (e.g., 0.736) is reached in a shorter time (cf. A and C).

**Table 7.1.** Examples of the influence of process parameters on logarithmic elution gradients

| Parameter[a] | Variants of the choice of parameters | | | |
|---|---|---|---|---|
| | A | B | C | D |
| $\varphi_{m,0}$[b] | 0 | 0 | 0 | 0 |
| $\varphi_{st}$ | 1 | 0.791 | 1 | 1 |
| $\dot{V}$ (ml/h) | 100 | 100 | 50 | 100 |
| $V_m$ (ml) | 300 | 300 | 300 | 600 |
| $t$ (h) | 0–8 | 0–8 | 0–8 | 0–8 |
| $\Delta$[c] | (0.487) | (0.385) | (0.284) | (0.284) |
| $\varphi_{m,2}$[b] | 0.487 | 0.385 | 0.284 | 0.284 |
| $\Delta$[c] | (0.249) | (0.197) | (0.203) | (0.203) |
| $\varphi_{m,4}$[b] | 0.736 | 0.582 | 0.487 | 0.487 |
| $\Delta$[c] | (0.129) | (0.102) | (0.145) | (0.145) |
| $\varphi_{m,6}$[b] | 0.865 | 0.684 | 0.632 | 0.632 |
| $\Delta$[c] | (0.066) | (0.052) | (0.104) | (0.104) |
| $\varphi_{m,8}$[b] | 0.931 | 0.736 | 0.736 | 0.736 |

[a] Symbols according to Eq. (7.1).
[b] $\varphi_{m,t}$ represents the solvent fraction in the elution mixture at $t = 0; 2; 4\ldots$ hours.
[c] $\Delta$ represents the difference in $\varphi_{m,t}$ between the foregoing and following $\varphi_{m,t}$ values. For a better orientation in the table, $\Delta$ values are written in parentheses.

- A bigger mixing vessel produces a flatter gradient (cf. A and D).
- Reduction of the flow rate or enlargement of the mixing volume influence the gradient in the same way (cf. C and D).
- Gradients having the same value $\varphi_{m,t}$ at a certain time (or at a certain elution volume, respectively) may have different courses depending on the combination of the individual parameters (cf. $\varphi_{m,8}$ and the values for the other times in B and D). That means that one can realize different gradients in a given period of operation which have the same starting and final values of the eluent ($\varphi_{m,0}$ and $\varphi_{m,t}$).

In practice, a compromise between the steepness of the gradient and the duration of elution time (or volumes of the solvent/nonsolvent mixture, respectively) is often necessary.

A big mixing vessel, which may be necessary by reason of either a very flat gradient or a voluminous column, requires a lot of nonsolvent mixture. Using a smaller mixing vessel saves solvent and nonsolvent and can yield the same gradient by a stepwise procedure (see Appendix A 9).

### 7.1.2 Temperature Programs

Increasing solvent power with constant composition of the eluent can be achieved by using temperature programs instead of solvent/nonsolvent gradients. A linear change of temperature with time is usually used [13].

Coupling of solvent/nonsolvent gradients with temporal or spatial temperature changes will be discussed in Sects. 7.2 and 8.

## 7.2 Principles and Limitations of Application

Gradient-elution fractionation in a column can be considered as a chromatographic process – this has already been discussed for the column-extraction procedure in Sect. 6.1. In gradient elution, however, not only the concentration but also the composition of the sole phase varies along the column (solvent/nonsolvent ratio and MWs of the eluted molecules).

Some peculiarities of the stationary gel phase must be noted: (i) The gel phase works only in the swollen state and develops from the action of the eluent. (ii) The gel phase alters continuously due to variations in the dissolution/precipitation equilibrium. (iii) The stationary phase descends in the course of the elution procedure and disappears finally.

Sol/gel equilibria in the column are directed by Eqs. (3.8) through (3.11). Pre-fractionation of the polymer in the formation of a gel phase is important especially for gradient elution (see point "Loading the support" in Sect. 6.4.2, p. 77).

Gradient-elution fractionations can be carried out in a normal as well as in a reversed-flow regime (cf. Fig. 6.2). Different solvent/nonsolvent gradients, discussed in Sect. 7.1.1, can be combined with temporal temperature programs. Mencer and Kunst [14] have superimposed a logarithmic solvent/nonsolvent gradient with a temporally linear decrease of the temperature to reach the best adjustment of the solubility gradient to the special sample. In this procedure the steepness of the gradient is lowered by the action of temperature. The following fractionations were carried out in this way:

PS/MEK/EtOH, 60 → 20 °C, and PS/CHN/PrOH, 70 → 20 °C. The choice of the conditions results from interpolation of solubility curves obtained by stepwise column extraction at different temperatures.

Other combinations of elution gradients and temperature changes are likewise possible. Such fractionation regimes should be tested in each special case.

**Optimization of Conditions** *Fractionation conditions* must be determined experimentally. The composition limits of the solvent/nonsolvent gradient can be determined easily via cloud-point or turbidimetric-titration experiments as mentioned on pp. 54 and 71. Optimization of the other parameters (Eq. (7.1), load, temperature, and others) is often required and must be performed by pilot fractionations with subsequent comparison of the fractionation efficiency (cf. Sects. 4.3.1 and 4.3.2). The careful choice of the experimental parameters is much more important in gradient elution than in stepwise column extraction.

Let us discuss some points which should be taken into consideration for the following general types of polymer samples:

*Low-MW samples* can be fractionated with a relatively steep gradient, but, a fine-grained support is advantageous.

*High-MW samples* require flat gradients. For that, poor solvents and nonsolvents and a relatively big mixing vessel should be used. As an example, Schneider et al. [15] fractionated PS samples with MW = $3 \times 10^5$ g/mol and $5 \times 10^6$ g/mol under otherwise equal conditions by use of mixing vessels of 220 ml and 2 l, respectively. Eventual adsorption on the support grows with increasing MW. The polymer concentration in the sol phase should decrease with growing MW (see Sect. 7.3).

Samples having *wide MWD* can be fractionated starting with a higher sample load. The most important matter is here a fairly adopted gradient.

Samples with *narrow MWD* require small and pre-fractionating loading the support, flatter gradients in comparison with a wide-MWD sample of similar MW range, slow flow rates, and relatively big mixing vessels.

Two problems, having minor importance in column extraction, should be stressed here for the gradient elution. *Overload* of the support diminishes markedly the fractionation efficiency and leads to reversals in the sequence of fractions. Besides, it is important that *dead volumes*, e.g., the volume between the glass frit and the fraction outlet in Fig. 6.2 (normal flow), are as small as possible to prevent the re-mixing of neighbouring fractions. This can be reached best by use of capillaries.

Summarizing this section, the following advantages and disadvantages can be stated.

**Advantages**

- The fractionation proceeds continuously and the period of fractionation is short in comparison with other fractionation procedures.
- Gradient-elution fractionation is suitable for automation.
- The method possesses a great variety for optimizing the fractionation conditions.
- Scaling-up of analytical procedures is possible.

**Disadvantages**

- Extensive preliminary investigations are required to fix the gradient and other parameters of the procedure.
- Loading the support is necessary before fractionation.
- The fractionation procedure *cannot be interrupted* after starting without causing disturbances!
- The ratio of total volume of solvent mixture and eluted polymer is relatively high due to the low polymer concentration in the sol phase.

Large-scale variants of the gradient-elution fractionation will be discussed together with Baker-Williams fractionation in Sect. 9.2.

## 7.3 Equipment and Materials

Most equipment for gradient elution is generally the same as used in column extraction. Thus, similar equipment can be used for degassing of the non-solvent and solvent mixtures, for loading the support, for collecting and processing of the fractions (see Sect. 6.4.1). This also applies to the column, i.e., the columns shown in Fig. 6.2 can generally be used in gradient elution.

An additional piece of apparatus is necessary for the *generation of the solvent/nonsolvent gradient*. A simple example for logarithmic gradients is shown in Fig. 7.1. The volumes of mixing and storage vessel must be chosen according to the demand of the gradient. Volumes of 200 to 500 ml for mixing vessel and about 2 or 3 l for storage vessel are usual for column length and diameter of about 1 m and 2.5 cm, respectively, and flow rates of about 100 ml/h. Smaller storage vessels can be used when the fractionation is closely supervised and the storage vessel refilled in time. In addition, a magnetic stirrer in the mixing vessel and a syphon between mixing and storage vessel are required. The syphon prevents the self-mixing of solvent and nonsolvent mixtures which is especially important for solvents having a higher density than the nonsolvent. The degasser should be situated above the syphon (see Fig. 7.1). A small stopcock in the uppermost region of the mixing vessel can be

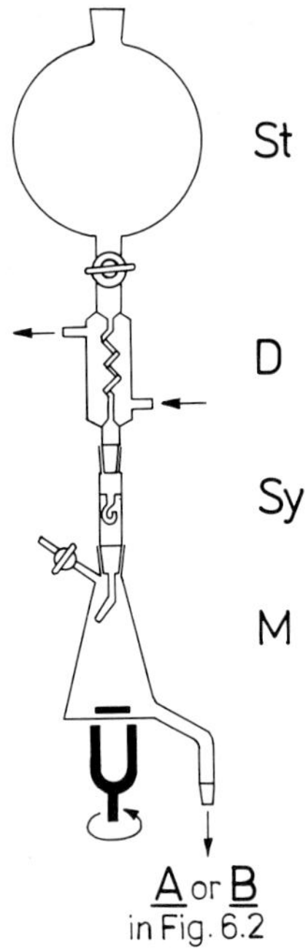


**Fig. 7.1.** Supplementary equipment to the column shown in Fig. 6.2 for the generation of logarithmic solvent/nonsolvent gradients. *St* – storage vessel, *D* – degasser, *Sy* – syphon, *M* – mixing vessel with stirrer. The shown equipment is arranged above the column according to A (normal flow) or B (reverse flow) in Fig. 6.2

useful to remove small gas volumes which, in case of improper degassing, may accumulate at the top of the vessel and reduce the mixing volume.

**Volumes, Gradients and Concentrations**

**Materials:** Support and solvent for loading can be estimated as in column extraction (see Sect. 6.4.1).

The volume of the starting *nonsolvent mixture* "A" is the sum of the interstitial volume inside the column and the volume of the mixing chamber including all connections.

The volume of the *solvent mixture* "B" in the storage vessel is directed by Eq. (7.1) ($\varphi_{st}$ and $\dot{V} \cdot t$, respectively). As in column extraction, some parameters influence the required eluent volume:

- The width of MWD determines the width (upper and lower limit) of the gradient and, consequently, the volume of the storage mixture "B" under certain conditions of flow rate and gradient steepness (cf. Sect. 7.1.1). Likewise, the composition of the mixture is influenced.
- The desired steepness of the gradient alters also the total volume and composition of the solvent mixture "B".
- The flow rate coupled with volume and cross-section of the column influences the required mixed solvent "B" in a certain elution time.
- The mass of polymer to be fractionated plays a role because the polymer concentration in the sol phase must not be too high. This problem directed the eluent volume necessary for the elution of a certain mass of polymer.

According to Flory [16] the concentration decreases with increasing MW:

$$c_{sol} \propto M^{-1/2} \tag{7.2}$$

The following rule of thumb may hold in most cases: The maximum polymer concentration in the sol phase (at middle MWs) should be about three times the average concentration which can be calculated from the polymer mass and the total volume of the eluent.

The highest possible sol-phase concentration depends on the fractionation system. Volume and composition of the mixed eluent "B" are involved among the other parameters of Eq. (7.1). Therefore, analysis of this equation is inalienable. Usually, the volume of the eluent (in ml) should be 200 to 500 times the mass of polymer to be fractionated (in g) using a higher ratio with increasing MW [5].

That means that about 5 g polymer can be fractionated in one fractionation run with about 1 l nonsolvent mixture "A" and 2 or 3 l solvent mixture "B" without any problems when a column with $l = 100$ cm and $d_i = 2.5$ cm, a flow rate $\dot{V} = 100$ ml/h, and a totally coated support are used.

For materials required for the processing of fractions see Sects. 4.2.2, 4.2.3, and p. 76.

## 7.4 Specific Steps of Implementation

### 7.4.1 Preparatory Investigations

Before the fractionation can start, some preliminary investigations and preparatory steps, similar to the column-extraction procedure (cf. Sect. 6.4.2), must be performed. The general choice of solvents, nonsolvents, and temperature should take into consideration the topics discussed in Sect. 4.1.2 and can be performed as described for precipitation and extraction fractionation (Sects. 5.3.1 and 6.5.2.1).

**Determination of Gradient Limits**

There are two possibilities to determine the limits of the solvent/nonsolvent mixture:

**Cloud-point measurements.**

- Solvent-rich mixture "B": The limit is found by addition of the *pure* nonsolvent to the polymer dissolved in the *pure* solvent at a suitable concentration (see below). The cloud point determines the composition of the end of the gradient ($\varphi_{m,t}$ for the final point, *not* the composition in the storage vessel, $\varphi_{st}$).
- Nonsolvent-rich mixture "A": The polymer is brought into contact with different solvent/nonsolvent mixtures of growing nonsolvent content. The mixture having least nonsolvent without any polymer in the supernatant

(what can be checked by addition of pure nonsolvent) yields the composition $\varphi_{m,0}$ for column and mixing vessel before the fractionation starts.

**Turbidimetric titration of the polymer dissolved in the *pure* solvent by addition of *pure* nonsolvent.**
The beginning and the end of the titration curve (cf. Fig. 5.4) correspond to the gradient limits just discussed.

These limits must be determined with a suitable concentration comparable with the real concentration conditions inside the column. For the starting mixture ("A") in the column and mixing vessel, a fictitious concentration (if the polymer were dissolved) can be calculated from the mass of polymer on the support and the (mixed) nonsolvent volume among the coated support. This condition can be used in the cloud-point method. In the turbidimetric titration, the polymer concentration in the final stage is very small, i.e., in the real case, if the polymer concentration had a higher fictitious value, all the polymer molecules would be precipitated. Thus, the end of the titration curve shows a solvent/nonsolvent composition that does not dissolve any portions of the polymer before the fractionation starts.

At the other limit (final point of fractionation, with mixture "B"), the polymer concentration is very small. Therefore, cloud-point determinations and turbidimetric titrations should be performed with low concentrations (e.g. 20–40 mg polymer per liter) and the influence of the concentration on the cloud point should be checked.

The determination of the gradient limits is often a matter of optimization.

Determination of the temporal temperature gradient by constant solvent/nonsolvent composition or by superposition with a solvent/nonsolvent gradient can be carried out analogously.

**Calculation of the Solvent/Nonsolvent Gradient, Starting and Storage Mixture, and Flow Rate**

Now one can use Eq. (7.1) with fixed values of $\varphi_{m,0}$ and $\varphi_{m,t}$ for the termination of the fractionation to calculate the other parameters and the course of the gradient ($\varphi_{m,t}$ for different $t$). Usually, one or two parameters are given by the experimental conditions (e.g., $V_m$ or $\dot{V}$). Dependent on the period of fractionation one can calculate the composition of the storage mixture. Note, however, the complex relations between the quantities of Eq. (7.1) shown in Table 7.1.

### 7.4.2 Fractionation Steps

**Preparation and Setting Up Equipment**

**Cleaning and loading the support.**
See column extraction, p. 77. A fine-grained support (appr. 0.1 mm diameter) should be used for low-MW samples (cf. Sect. 7.2).

**Filling the column.**
See column extraction, p. 78.

**Filling the mixing vessel and the syphon.**
The mixing vessel must be filled completely with *degassed* nonsolvent mixture "A" (as the column). The syphon must also be filled at the beginning with the nonsolvent mixture in the following way: The syphon is put upside down into a glass beaker which is filled with the non-solvent mixture (as high as the syphon is long). The syphon is filled with the nonsolvent mixture from the bottom, and the opening is then closed by inserting a ground-glass stopper which is at the bottom of the beaker. The filled syphon can now be inserted into the mixing vessel and the stopper can be removed. Then, the upper part of the syphon is emptied to the middle of the crook by letting the corresponding volume of mixture "A" flow out of the column. Finally, the empty volume of the syphon is filled with solvent mixture "B" (from the storage vessel).

**Filling the degasser and the storage vessel.**
Both must be filled with the solvent mixture "B" with composition $\varphi_{st}$.

**Thermostatting.**
See column extraction, p. 78. If the elution proceeds under continuous temperature change, a suitable device for temperature programming must be connected to the column jacket.

**Degassing.**
See column extraction, p. 79.

**Elution Procedure**

Problems like maintaining a constant temperature (when no temperature change takes place) and the amount of load are similar to extraction fractionation (Sect. 6.4.2). The flow rate, however, in the gradient technique, is of great importance because a consistently equal flow rate of the elution mixture within the cross-section is a prerequisite for a successful fractionation. A too-rapid flow rate may disturb the flow profile, i.e., the homogeneity of eluent composition in a certain cross-section of the column, and may cause chanelling. A flow rate that is too slow may lead to re-mixing of separated components in neighbouring cross-sections of the column caused by diffusion.

Owing to different compositions of the sol-phase volumes, the viscosity of the eluate changes (usually increases). Therefore, the flow rate should be monitored and, if necessary, readjusted. If elution is incomplete, pure solvent can be poured into the storage vessel *in the final stage* of the fractionation to dissolve all polymer components. (Keep in mind, however, that the elution gradient must be flat for the last and highest-MW components!)

**Collection and Processing of Fractions**

See column extraction, p. 79 and Sects. 4.2.2 and 4.2.3. If one wants to control the composition of the sol fractions, measurements of the refractive indices of the mixed eluate can yield information when the differences in refractive indices of the eluent components are great enough. The polymer concentration can be qualitatively checked by introducing the sol phase drop by drop into a strong precipitant.

If the dried fractions are very small, adsorption of steam or other substances on the surfaces of the drying vessels can falsify the masses of the fractions. Therefore, some additional vessels free of polymer should be treated in exactly the same way (washing, drying, weighing out) to correct the weights of the vessels containing polymer.

**Purging the Column** See column extraction, p. 80.

Problems of fractionation efficiency and trouble shooting will be discussed together with column-extraction and Baker-Williams procedures in Sect. 9.1.

## 7.5 Examples of Gradient-Elution Fractionations

Two gradient-elution fractionations of a styrene/acrylonitrile (S/AN) copolymer will be described in this section. These examples illustrate the fractionation of copolymers as discussed in Sect. 4.4. The copolymer is fractionated with respect to MW and chemical composition using dichloromethane/methanol and dichloromethane/*n*-hexane, respectively.

### 7.5.1 Gradient-Elution Fractionation of Poly(styrene-*co*-acrylonitrile) with Dichloromethane/Methanol

This example is suggested for different scales of columns and polymer masses. Table 7.2 summarizes some operating variables in dependence on column dimensions. Other values can be interpolated. See also Sects. 7.3 and 7.4.

**Materials** **Polymer.** Poly(styrene-*co*-acrylonitrile) (S/AN) of nearly azeotropic (mole fraction AN = 0.376) composition and $\bar{M}_n$ about $8 \times 10^4$ g/mol; masses see Table 7.2

**Solvent.** Dichloromethane (DCM) (analytical grade), 50 ml/1 g polymer for coating the support. Volume necessary for fractionation depends on column dimensions and characteristic of the gradient (see Table 7.2)

**Nonsolvent.** Methanol (analytical grade). The same criteria as for DCM hold in regard to the volume required

**Support.** Glass beads or sea sand ($d = 0.2$–$0.5$ mm), about 50 $cm^3$/1 g polymer. Calculations were performed for glass beads with diameter of 0.4 mm and support load of 22.2 mg/$cm^3$ (see Table 7.2)

Solvent and nonsolvent for precipitation, re-dissolution, and re-precipitation of the fractions must be related to their masses. Approximate quantities: $\leqslant$100 ml solvent/1 g polymer (re-dissolution), $\geqslant$ 1 l nonsolvent/1 g polymer

**Table 7.2.** Operating variables of gradient elution (Sect. 7.5.1)

| Column dimensions: $l$ (cm) | 20 | 50 | 100 | 50 | 100 |
|---|---|---|---|---|---|
| $d_i$ (cm) | 2.5 | 2.5 | 2.5 | 7.0 | 7.0 |
| Support (bulk volume) ($cm^3$) | 100 | 245 | 490 | 1900 | 3800 |
| Polymer[a] (g) | 2.2 | 5.4 | 10.8 | 42 | 84 |
| Interstitial volume (appr.) (ml) | 26 | 64 | 128 | 495 | 990 |
| Flow rate, $\dot{V}$ (ml/min) | 2 | 2 | 2 | 17 | 17 |
| Mixing vessel, $V_m$ (l) | 0.3 | 0.3 | 0.3 | 2.5[d] | 2.5[d] |
| $V$ (nonsolvent mixture) (ml) | 400 | 450 | 500 | 900 | 1400 |
| $V$ (DCM)[b] (ml) | 56 | 63 | 70 | 126 | 196 |
| $V$ (MeOH)[b] (ml) | 344 | 387 | 430 | 774 | 1205 |
| $V$ (solvent mixture)[b,c] (ml) | 1240 | 1290 | 1340 | 8040 | 8540 |
| $V$ (DCM)[b] (ml) | 805 | 837 | 870 | 5218 | 5542 |
| $V$ (MeOH)[b] (ml) | 435 | 453 | 470 | 2822 | 2998 |
| $\sum V$ (DCM) (ml) | 861 | 900 | 940 | 5345 | 5740 |
| $\sum V$ (MeOH) (ml) | 779 | 840 | 900 | 3596 | 4202 |
| Fractionation time, $t$[b] (h) | 7 | 7 | 7 | 7 | 7 |

[a] Values are related to glass beads of diameter 0.4 mm and a load (polymer/bulk volume) of 22.2 mg/cm$^3$ (=12.1 mg/g = 0.2 mg/cm$^2$).
[b] Values are calculated according to Eq. (7.1) assuming a gradient in the range of $\varphi_s = 0.140$–$0.607$ (see Fig. 7.2) and using $\varphi_{m,0} = 0.140$, $\varphi_{st} = 0.637$, $\varphi_{m,7} = 0.607$, and $\dot{V}$ and $V_m$ as given above. Deviations of any of these parameters yield other values.
[c] The sum of interstitial volume and $V_m$ (nonsolvent mixture) must be added to the product $\dot{V} \cdot t$.
[d] Use of a smaller mixing vessel yielding the same gradient is described in the Appendix (A 9).

(precipitation or re-precipitation). Solvents for support cleaning are not considered.

**Equipment**

- Glass beaker (250 ml) for coating the support. Greater volumes of support should be coated in several portions with identical polymer/solvent/support ratio.
- Glass rod
- Ventilator and exhauster
- Mortar
- Column as shown in Fig. 6.2 furnished with thermo-jacket, recirculating thermostat, and valve. Use in normal-flow direction. Column dimensions are given in Table 7.2
- Mixing vessel with stirrer, syphon, degasser, and storage vessel (2 l) according to Fig. 7.1
- Collecting vessels for fractions. Number and volumes depend on number and volume of sol fractions. A fraction collector may be useful. (For an analytic fractionation yielding many small fractions, a set of small glass beakers, including a number of extra beakers – approx. 30% (see Fractionation) can be used.)

- Rotational-type evaporator for concentration of the sol phases (Small fractions in beakers can be carefully dried directly using a slight vacuum in an oven.)
- Glass beaker for precipitation or re-precipitation of fractions (>1 l/1 g polymer)
- Stirrer
- Dropping funnel
- Device for filtration under vacuum or pressure (filter from sintered glass or paper)
- Shallow glass dishes
- Vacuum oven
- Stands, clamps, sockets, etc.

**Time Required**

- Dissolution of polymer 2–3 h
- Coating the support (one portion) 2–3 h
- Filling and assembly of column, mixing vessel, syphon, and storage vessel 1 h
- Thermostatting 1 h
- Elution of the polymer 7 h
- Purging of the packed column 0.5–2 h

Times necessary for cleaning the support and processing of the fractions are not included.

**Preparatory Investigations**

- Determination of the gradient limits at fractionation temperature and $c_P = 50$ mg/l by turbidimetric titration (cf. Sect. 7.4.1). The result of an assumed S/AN sample is shown in Fig. 7.2 taking into consideration reserves to ensure total solubility and insolubility, respectively.
- Calculation of the gradient parameters using Eq. (7.1), the determined limits $\varphi_{m,0}$ and $\varphi_{m,t}$ (including small reserves), $\dot{V}$, and $V_m$ according to the

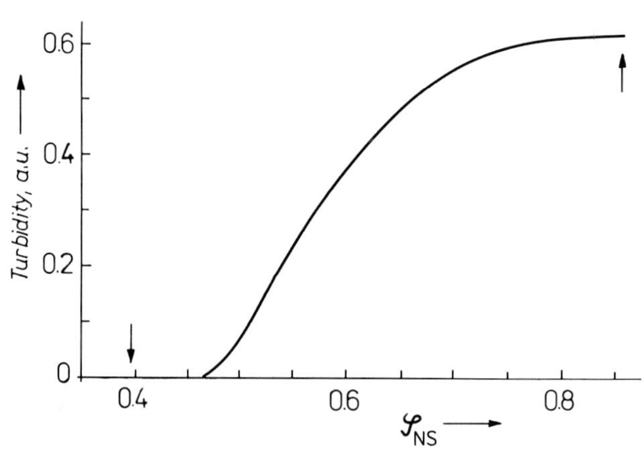


**Fig. 7.2.** Assumed turbidimetric-titration curve of the S/AN sample fractionated in the example in Sect. 7.5.1 (solvent dichloromethane, nonsolvent methanol, $T = 20\,°C$). The *arrows* mark the limits of the gradient used in the gradient-elution fractionation

apparatus. Now, $\varphi_{st}$ can be adjusted relative to the desired fractionation time. For example, with $\varphi_{m,0} = 0.14$, $\varphi_{m,7} = 0.607$ (related to the solvent component and to a fractionation time $t = 7$ h; cf. Fig. 7.2 where $\varphi_{NS}$ is used), $V_m = 300$ ml, and $\dot{V} = 120$ ml/h, $\varphi_{st}$ can be calculated by rearrangement of Eq. (7.1):

$$\varphi_{st} = [\varphi_{m,7} - \varphi_{m,0} \exp(-\dot{V} \cdot t/V_m)]/[1 - \exp(-\dot{V} \cdot t/V_m)] = 0.637\,.$$

With Eq. (7.1) it is now possible to calculate $\varphi_{m,t}$, e.g., $\varphi_{m,3} = 0.487$ results after 3 h. For deviating gradient limits, one calculates parameters which are different from those given in Table 7.2. Please, calculate the course of your gradient for different periods of time!

- If the gradient shall be produced with help of a smaller mixing vessel (cf. p. 107), consult Appendix A 9.

**Fractionation**

Fractionation temperature: 20 °C

- Cleaning the support by washing with concentrated hydrochloric acid (exhauster!), distilled water (repeatedly), and DCM
- Dissolution of the polymer in DCM, concentration about 2–5%
- Addition of the corresponding amount of support (see Table 7.2) to the polymer solution, slight heating with stirring by hand and evaporation of the solvent (ventilator and exhauster)
- Careful crushing of clumps; quantitative removal of the coated support from the beaker
- Filling the column (after the outlet is closed) with degassed nonsolvent mixture according to the interstitial volume (see Table 7.2) and afterwards with the coated support. The support must be kept covered with liquid.
- Setting up the column (Fig. 6.2, normal flow) and mixing vessel (Fig. 7.1)
- Filling the mixing vessel with degassed nonsolvent mixture
- Filling the syphon with degassed nonsolvent mixture, see p. 113. The syphon must be filled so that it contains no air.
- Filling the degasser and the storage vessel with solvent mixture. Refilling the storage vessel may be necessary.
- Thermostatting of column and degasser
- Start of the elution by opening the foot valve of the column and regulation of the desired flow rate. Flow rate must be monitored and adjusted continuously.
- Collection of the sol fractions in suitable flasks or beakers depending on number and mass of the fractions. Generally, the eluate can be divided into as many fractions as desired.
- Elution of the column with a plenty of pure solvent or at some higher temperature after the last fraction (purging of support and column)
- Processing of the fractions by either rotational-type evaporation and precipitation or direct drying in a vacuum. The latter procedure is advantageous when small fractions (< 50 mg) are collected. In this case, empty

but identically treated (washed, dried, weighed) beakers should also be weighed to eliminate the effect of possible adsorption on the glass surfaces of the beakers.

**Evaluation**

- Determination of mass and MW or related values ([$\eta$], $\varphi^*$, $V_e$ from SEC) for each fraction – gradation should be found
- Control of CC of the fractions (e.g., by Kjeldahl analysis) – no gradation should be found
- Comparison of gradation and sharpness of fractions and starting polymer by SEC or turbidimetric titration (are inversions found?)
- Comparison of directly measured and calculated averages of MW or [$\eta$] and compositions for the whole sample according to Eqs. (2.2) to (2.4) and (4.1).

Conclusions about limits, course (steepness), and parameters of the solvent/nonsolvent gradient should be deduced to optimize the fractionation (see Sects. 7.1.1 and 9.1).

### 7.5.2 Gradient-Elution Fractionation of Poly(styrene-*co*-acrylonitrile) with Dichloromethane/*n*-Hexane

This fractionation is performed as in Sect. 7.5.1 leading to fractions graded with respect to AN content. Table 7.3 summarizes operating variables analogously to Table 7.2. See also Sects. 7.3 and 7.4.

**Materials**

**Polymer.** Poly(styrene-*co*-acrylonitrile) (S/AN) of nearly azeotropic (mole fraction AN = 0.376) composition and $\bar{M}_n$ about $8 \times 10^4$ g/mol; masses see Table 7.3

**Solvent.** Dichloromethane (DCM) (analytical grade), 50 ml/1 g polymer for coating the support. Volume necessary for fractionation depends on column dimensions and characteristic of the gradient (see Table 7.3)

**Nonsolvent.** *n*-hexane (analytical grade). The same criteria for DCM hold in regard to the volume required

**Support.** Glass beads or sea sand ($d$ = 0.2–0.5 mm), about 50 cm$^3$/1 g polymer. Calculations were performed for glass beads with diameter of 0.4 mm and support load of 22.2 mg/cm$^3$ (see Table 7.3)

Solvent and nonsolvent for precipitation, re-dissolution, and re-precipitation of the fractions must be related to their masses. Approximate quantities: $\leqslant$100 ml solvent/1 g polymer (re-dissolution), $\geqslant$1 l nonsolvent/1 g polymer (precipitation or re-precipitation). Solvents for support cleaning are not considered.

**Table 7.3.** Operating variables of gradient elution (Sect. 7.5.2)

| Column dimensions: $l$ (cm) | 20 | 50 | 100 | 50 | 100 |
|---|---|---|---|---|---|
| $d_i$ (cm) | 2.5 | 2.5 | 2.5 | 7.0 | 7.0 |
| Support (bulk volume) ($cm^3$) | 100 | 245 | 490 | 1900 | 3800 |
| Polymer[a] (g) | 2.2 | 5.4 | 10.8 | 42 | 84 |
| Interstitial volume (appr.) (ml) | 26 | 64 | 128 | 495 | 990 |
| Flow rate, $\dot{V}$ (ml/min) | 2 | 2 | 2 | 17 | 17 |
| Mixing vessel, $V_m$ (l) | 0.3 | 0.3 | 0.3 | 2.5[d] | 2.5[d] |
| $V$ (nonsolvent mixture) (ml) | 400 | 450 | 500 | 900 | 1400 |
| $V$ (DCM)[b] (ml) | 192 | 216 | 240 | 432 | 672 |
| $V$ (Hx)[b] (ml) | 208 | 234 | 260 | 468 | 728 |
| $V$ (solvent mixture)[b,c] (ml) | 1240 | 1290 | 1340 | 8040 | 8540 |
| $V$ (DCM)[b] (ml) | 770 | 801 | 832 | 4993 | 5303 |
| $V$ (Hx)[b] (ml) | 470 | 489 | 508 | 3047 | 3237 |
| $\sum V$ (DCM) (ml) | 962 | 1017 | 1072 | 5425 | 5975 |
| $\sum V$ (Hx) (ml) | 678 | 723 | 768 | 3515 | 3965 |
| Fractionation time, $t$[b] (h) | 7 | 7 | 7 | 7 | 7 |

[a] Values are related to glass beads of diameter 0.4 mm and a load (polymer/bulk volume) of 22.2 mg/cm$^3$ (= 12.1 mg/g = 0.2 mg/cm$^2$).
[b] Values are calculated according to Eq. (7.1) assuming a gradient in the range $\varphi_s = 0.480$–$0.610$ and using $\varphi_{m,0} = 0.480$, $\varphi_{st} = 0.618$, $\varphi_{m,7} = 0.610$, and $\dot{V}$ and $V_m$ as given above. Deviations of any of these parameters yield other values.
[c] The sum of interstitial volume and $V_m$ (nonsolvent mixture) must be added to the product $\dot{V} \cdot t$.
[d] Use of a smaller mixing vessel yielding the same gradient is described in the Appendix (A 9).

**Equipment**

The gradient-elution fractionation with dichloromethane and *n*-hexane can be carried out using the same equipment as described in Sect. 7.5.1. Table 7.3 is now valid instead of Table 7.2.

**Time Required**

The experimental conditions according to Table 7.3 make possible the fractionation in the same period as required in Sect. 7.5.1.

**Preparatory Investigations**

Generally, the preparatory investigations follow Sect. 7.5.1. The fractionation with dichloromethane/*n*-hexane is, however, not illustrated by a turbidimetric-titration curve. The example of the gradient calculation given in Sect. 7.5.1 must now be adapted using the actual values specified in Table 7.3.

**Fractionation**

The course of the fractionation is analogous to the gradient elution described in Sect. 7.5.1. Specific conditions can be seen in Table 7.3.

**Evaluation** The points of view from Expl. 7.5.1 are valid, however, gradation should be found not for MW but for AN content. Evaluations can be made additionally for CC, i.e.,

- bar graph vs AN content and
- distribution curves $I(w_{AN})$ and $W(w_{AN})$.

## References


1. Desreux V (1949) Rec Trav Chim 68: 1789; Desreux V, Spiegels MC (1950) Bull Soc Chim Belges 59: 476; Desreux V, Oth A (1952) Chem Weekblad 48: 247
2. Glöckner G (1987) Polymer characterization by liquid chromatography. Elsevier, Amsterdam
3. Schröder E, Müller G, Arndt KF (1988) Polymer characterization. Hanser, Munich; (1989) Polymer characterization. Akademie-Verlag, Berlin
4. Elliot JH (1967) In: Cantow MJR (ed) Polymer fractionation. Academic, New York, p 67
5. Porter SR, Johnson JF (1967) In: Cantow MJR (ed) Polymer fractionation. Academic, New York, p 95
6. Barrall EM, Johnson JF, Cooper AR (1977) In: Tung LH (ed) Fractionation of synthetic polymers. Marcel Dekker, New York, p 267
7. Guillet JE, Combs RL, Slonaker DF, Coover Jr HW (1960) J Polym Sci 47: 307
8. Meyerhoff G, Romatowski J (1964) Makromol Chem 74: 222
9. Donaldson KO, Tulane VJ, Marshall LM (1952) Anal Chem 24: 185
10. Bock RM, Ling NS (1954) Anal Chem 26: 1543
11. Gernert JF, Cantow MJR, Porter RS, Johnson JF (1963) J Polym Sci C 1: 195
12. Mikeš O, Vespalec R (1975) In: Deyl Z, Macek K, Janák J (eds) Liquid column chromatography. A survey of modern techniques and applications. Elsevier, Amsterdam, p 233
13. Klein J, Friedel H (1970) J Appl Polym Sci 14: 1927
14. Mencer HJ, Kunst B (1978) Colloid Polym Sci 256: 696; (1979) Makromol Chem 180: 2463
15. Schneider NS, Laconti JD, Holmes LG (1960) J Appl Polym Sci 3: 251
16. Flory PJ (1953) Principles of polymer chemistry. Cornell University Press, Ithaca, New York, p 341

# 8 Baker-Williams Fractionation

## 8.1 Principles and Limitations of Application

The precipitation chromatography method suggested by Baker and Williams [1] is the combination of gradient elution (cf. Sect. 7) with an antiparallel temperature gradient [2–5]. Supports coated with the polymer are the same as in gradient elution, but only 5 to 10% of the support may be loaded with polymer. This support fraction is arranged at the hottest part of the column, i.e., usually at the top (when the eluent works in the normal-flow direction). The rest of the support is available for the chromatographic process. The solvent/nonsolvent gradient operates in the same way as in gradient elution.

The *temperature gradient* along the column is usually linear produced by heating the top and cooling the bottom of the column (see Sect. 8.2).

Two other variants of fractionation according to the principle of Baker and Williams are known: (i) Use of a single theta-solvent instead of the solvent/nonsolvent gradient in combination with programmed heating of the whole column with maintenance of the constant, spatial temperature gradient [6], and (ii) use of a common elution gradient together with a pulsed temperature program, i.e., periodically fast heating and slow cooling of the whole column [7]. Each cycle of the program must be shorter than the period necessary for the flow of any eluate volume through the column.

With the solvent/nonsolvent gradient, molecules with a certain MW are eluted from the support at the highest temperature and migrate in zones of lower temperature leading to phase separation and coating of the support. The following solvent/nonsolvent mixture with increased dissolution power can now dissolve this polymer and transports it to the next cooler zone to produce a new gel phase at the support. Thus, a continuous exchange between the stationary but permanently altered gel phase and the moving sol phase takes place according to temperature and eluent mixture at a certain locus and at a certain time. When the gel phase forms droplets instead of a film inside the column, the chromatographic process will be disturbed or even impossible.

Precipitation chromatography can be described by a theory by Schulz et al. [8], which says that the high efficiency of Baker-Williams fractionation is due to the combination of (i) an elution effect, (ii) a compression effect, and (iii) a migration effect.

The width of the temperature gradient must be tested in each case. General limits come from the boiling and freezing points of the solvent and nonsolvent.

The most common differences in upper and lower temperatures are about 40 K for column lengths between about 35 and 100 cm. Other temperature differences are also used (see [2, 5]). Fractionation power is lost when the temperature gradient is too small. A gradient which is too high can possibly cause the fractionation efficiency to deteriorate when the heat conductivity across the column is too low. Then chanelling occurs with a higher flow rate of the polymer in the center of the packing and a lower one at the periphery of the column.

Concerning the *solvent/nonsolvent gradient*, all statements given for gradient elution (cf. Sect. 7) hold true for precipitation chromatography as well. However, the limits of the gradient must now be determined for the temperature at the top and at the bottom of the column. Complete precipitation must be reached at the top temperature and the cloud point must be measured at the bottom temperature.

Finally, advantages and disadvantages of Baker-Williams fractionation are summarized. (in addition to those mentioned for gradient elution):

**Advantages**

- Sharper fractions (optimum conditions provided)
- Broader variety of possibilities to optimize fractionation conditions.

**Disadvantages**

- More extensive preparatory tests
- Additional temperature monitoring
- Precautions against chanelling necessary.

Surveys of Baker-Williams fractionations reported in the literature are given in Refs. [2, 3, 5].

## 8.2 Equipment and Materials

The equipment for precipitation chromatography is generally the same as for gradient-elution fractionation (see Sect. 7.3). However, a device for *generation of the temperature gradient* is additionally required. There are different possibilities for that [2]:

**Temperature Gradient Generating Devices**

Usually, a glass column is inserted in a cylindrical metal block whose one end is heated and the other end is thermostatted at a lower temperature. A linear temperature gradient develops with good insulation and a metal of high heat conductivity. (For this reason, the jacket is usually made of aluminum.) In a similar way, compact columns from metal with thick walls containing the heating and cooling devices at the ends can be used. Effective insulation is necessary. Gold-plated surfaces prevent any catalytic influence of the metal on the polymers.

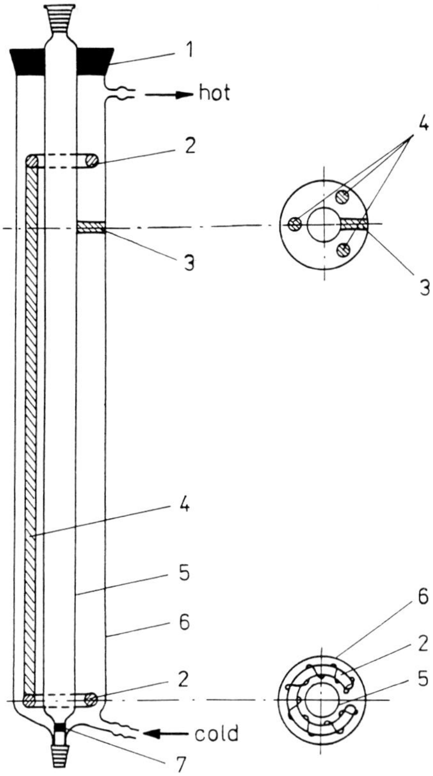


**Fig. 8.1.** All-glass column for generation of temperature gradients (schematically). *1* – rubber stopper, *2* – support for heating wires, *3* – stabilizer between inner and outer tube, *4* – glass prop, *5* – fractionation tube, *6* – water jacket, *7* – glass frit *G1*; heating wires clamped parallel between the supports *2* are omitted (Ref. [9] with permission of Acta Polymerica)

A temperature gradient can also be generated by use of a coil of thermowire wrapped around a glass column with varying distances between the windings (smaller distances give hotter zones). Good insulation must be ensured.

Another variant is the use of a column manufactured completely from glass [9]. This column is shown in Fig. 8.1. The temperature gradient is produced in a glass jacket equipped with vertically oriented and parallelly connected electrical heating wires where water flows slowly in an upward direction. The constantly flowing water is heated along the column and leads to a linear gradient. Details of the water circuit can be taken from preparative all-glass column shown in Fig. 9.1. Such a column has been successfully used with the following parameters: length = 100 cm, diameter of the column = 2.4 cm, diameter of the jacket = 7 cm, 12 wires with a resistance of 78 Ohm/m, flow rate of water about 8 l/h, heating voltage = 25 V (only *low voltages should be used* for safety reasons!). Flow rate and amperage depend on the temperature difference and must be experimentally tested. Gradients of 40 K per 100 cm were possible without problems.

The temperature gradient must cover only about 90% of the column length since the sample bed (about 10% of the support) must have a constant (hot) temperature (cf. Fig. 8.1).

**Temperature Profile** An important role in Baker-Williams fractionation is played by the *temperature profile* in the cross sections of the column bed. The temperature in a certain cross section should be constant in the idealized case. However, owing to the flow of the sol phase from zones of higher temperature to lower temperature ones, it takes longer for the flowing sol phase to reach the temperature of the column wall the further it is from the wall. Thus, the profile of constant temperature sags in the middle of the cross section. This fact limits the column diameter to usually less than 4 cm (see Sects. 9.1 and 9.2). In addition, the temperature gradient must not be too steep and the flow rate not too high.

**Materials** Concerning *materials* required for precipitation chromatography, the same estimations as for gradient elutions can be made (cf. Sect. 7.3). Note, however, that generally only a small portion of the support is coated by the polymer in Baker-Williams fractionation.

## 8.3 Specific Steps of Implementation

### 8.3.1 Preparatory Investigations

Preparatory steps are very similar to gradient-elution fractlionation – see therefore Sect. 7.4.1.

**Choice of the Temperature Gradient** No reliable rules exist for this problem. The optimum must be found by experimental comparison and may depend on the boiling points of the components. A temperature difference of 40 to 50 K is often used, but, it must be pointed out that a temperature gradient which is too steep involves a steeper solvent gradient. The latter adversely affects the fractionation efficiency, especially for high MWs (cf. pp. 108 and 113), because the first step in precipitation chromatography is the elution of the polymer out of the coated support.

**Solvent/Nonsolvent Gradient** Tests like those in Sect. 7.4.1 must be related to total solubility at the *lowest* temperature and complete insolubility at the *highest* temperature. Calculation of gradient, starting and storage mixtures, and flow rate according to Eq. (7.1) can be performed as described in Sect. 7.4.1 and Table 7.1. For the influence of the flow rate see also below.

### 8.3.2 Fractionation Steps

**Cleaning and loading the support.**
See column extraction, p. 77.

**Preparation and Setting Up Equipment**

**Degassing.**
See column extraction, p. 79.
The temperature of degassing of the eluent mixture must lie above the top temperature of the column. The nonsolvent-rich mixture "A" as well as mixture "B" must be degassed before use.

**Filling the column.**
See column extraction, p. 78.
In addition, it should be noted that the sample bed (zone of coated support, 5 to 10% of the column) must be above the temperature-gradient region because the elution step must occur at constant top temperature.

**Filling the mixing vessel, the syphon, the degasser, and the storage vessel.**
See gradient-elution fractionation, p. 113.

**Generation of the temperature gradient.**
Temperatures at bottom and top of the column should be continously monitored. The whole column must have reached perfect temperature equilibrium according to the gradient before the elution can be started.

**Elution Procedure**

See gradient-elution fractionation, p. 113.

Two points concerning the flow rate should be emphasized: (i) The faster the flow rate the greater is the sagging of the temperature profile over the cross section of the column (cf. Sect. 8.2). (ii) A flow rate which is too slow promotes the back-diffusion of polymer molecules into zones of better solubility which leads to re-mixing of fractions.

**Collection and Processing of Fractions**

See column extraction, p. 79, and gradient-elution fractionation, p. 113.

**Purging the Column**

See column extraction, p. 80.

Problems of fractionation efficiency and trouble shooting will be discussed together with column-extraction and gradient-elution procedures in Sect. 9.1.

## 8.4 Example for a Baker-Williams Fractionation of Poly(styrene) with Toluene/Methanol

Table 8.1 gives a survey of operating conditions.

**Materials**

**Polymer.** Poly(styrene), $\bar{M}_n < 10^5$ g/mol, $\bar{M}_w$ about $2 \times 10^5$ g/mol; masses see Table 8.1

**Solvents.** Toluene (analytical grade, used for fractionation), approximated volumes see Table 8.1;

**Table 8.1.** Operating conditions of Baker-Williams fractionation (Sect. 8.4)

| | |
|---|---|
| Column dimensions: $l$ (cm) | 100 |
| $d_i$ (cm) | 2.5 |
| Support (bulk volume) ($cm^3$) | 490 |
| Coated part (5–10%) ($cm^3$) | 24–49 |
| Polymer[a] (g) | 0.55–1.1 |
| Interstitial volume (appr.) (ml) | 130 |
| Flow rate, $\dot{V}$ (ml/min) | 2 |
| Mixing vessel, $V_m$ (l) | 0.3 |
| Temperature gradient: $T_{foot}$ (°C) | 15 |
| $T_{top}$ (°C) | 50 |
| $V$ (nonsolvent mixture) (ml) | 500 |
| $V$ (T)[b] (ml) | 250 |
| $V$ (MeOH)[b] (ml) | 250 |
| $V$ (solvent mixture)[b, c] (ml) | 1340 |
| $V$ (T)[b] (ml) | 1100 |
| $V$ (MeOH)[b] (ml) | 240 |
| $\sum V$ (T) (ml) | 1350 |
| $\sum V$ (MeOH) (ml) | 490 |
| Fractionation time, $t$[b] (h) | 7 |

[a] Values are related to glass beads of diameter 0.4 mm and a load (polymer/bulk volume) of 22.2 mg/cm$^3$ (= 12.1 mg/g = 0.2 mg/cm$^2$).
[b] Values are calculated according to Eq. (7.1) assuming a gradient in the range $\varphi_s = 0.50$–$0.80$ and using $\varphi_{m,0} = 0.50$, $\varphi_{st} = 0.82$, $\varphi_{m,7} = 0.80$, and $\dot{V}$ and $V_m$ as given above. Deviations of any of these parameters yields other values.
[c] The sum of interstitial volume and $V_m$ (nonsolvent mixture) must be added to the product $\dot{V} \cdot t$.

Dichloromethane (DCM) (analytical grade; used for coating the support), 40–50 ml/1 g polymer

**Nonsolvent.** Methanol (analytical grade), approximated volumes see Table 8.1

**Support.** Glass beads or sea sand ($d = 0.2$–$0.5$ mm), about 50 cm$^3$/1 g polymer. Calculations were performed for glass beads with diameter of 0.4 mm and support load of 22.2 mg/cm$^3$ (see Table 8.1).

Solvent and nonsolvent for precipitation, re-dissolution, and re-precipitation of the fractions must be related to their masses. Approximate quantities: $\leqslant 100$ ml solvent/1 g polymer (re-dissolution), $\geqslant 1$ l nonsolvent/1 g polymer (precipitation or re-precipitation). Solvents for support cleaning are not considered.

**Equipment**

- Glass beaker (100 ml) for coating the support
- Glass rod
- Ventilator and exhauster
- Mortar
- Analytic all-glass column described in Sect. 8.2 and shown in Fig. 8.1 is recommended.[1] Dimensions are given in Table 8.1 (possible deviations influence other parameters in the table). Water jacket is equipped with parallel heating wires.
- Devices to produce and to control the temperature gradient:

• Large recirculating thermostat equipped with an overflow and connected with the external water circuit
• Vessel with overflow connected to the thermostat and foot inlet of the water jacket. The overflow leads back to the thermostat. The vessel is connected so that it is above the top of the column to facilitate a constant and smooth flow through the jacket.
• Cooling device connected at the top of the water jacket and the thermostat in order to cool the hot water flowing back to the thermostat
• Thermometers before the water inlet and after the outlet to monitor the temperature at the foot and top of the column
• *Low tension* source with adjustable power output up to about 1000 W to heat the wires. It must be possible to check the power output by measuring the temperature inside the water outlet of the column jacket.
• Valves to adjust the water flow through the jacket
• Flow-meter for better control of the water flow

- Mixing vessel with stirrer, syphon, degasser, and storage vessel (2 l) according to Fig. 7.1
- Collecting vessels for fractions. Number and volumes depend on number and volume of sol fractions. A fraction collector may be useful. (For an analytic fractionation yielding many small fractions, a set of small glass beakers, including 30% extra beakers (see Fractionation) can be used.)
- Rotational-type evaporator for concentrating the sol phases (Small fractions in beakers can be carefully dried directly using a slight vacuum in an oven.)
- Glass beaker for precipitation or re-precipitation of fractions (>1 l/1 g polymer)
- Stirrer
- Dropping funnel
- Device for filtration under vacuum or pressure (filter of sintered glass or paper)

[1] Here, relatively simple column equipment is shown. Additional devices for control of flows and temperatures can be used.

- Shallow glass dishes
- Vacuum oven
- Stands, clamps, sockets, etc.

**Time Required**

- Dissolution of polymer 2–3 h
- Coating the support 2–3 h
- Filling and assembly of column, mixing vessel, syphon, and storage vessel 1 h
- Adjustment of the temperature gradient 1.5 h
- Elution of the polymer 7 h
- Purging of the packed column 0.5–2 h

Times necessary for cleaning the support and processing of the fractions are not included.

**Preparatory Investigations**

- Determination of the solvent/nonsolvent gradient limits by turbidimetric titration of solutions with $c_p = 50$ mg/l at 15 and 50 °C. The cloud point (limit of complete solubility) must be determined at 15 °C whereas the point of total precipitation must be measured at 50 °C. With the PS used for fractionation, values $\varphi_{m,0}$ and $\varphi_{m,t}$ (Eq. (7.1)) are expected to deviate slightly from the values assumed in Table 8.1.
- Calculation of the gradient parameters using Eq. (7.1), the determined limits $\varphi_{m,o}$ and $\varphi_{m,t}$ (including small reserves), $\dot{V}$, and $V_m$ according to the apparatus. Now, $\varphi_{st}$ can be adjusted relative to the desired fractionation time. For deviating gradient limits, one calculates parameters which are different from those given in Table 8.1. Please, calculate the course of your gradient for different periods of time (see Sect. 7.5.1, Paragraph on preparatory investigations).

**Fractionation**

The water circuit should be already installed.

- Cleaning the support by washing with concentrated hydrochloric acid (exhauster!), distilled water (repeatedly), and DCM
- Dissolution of the polymer in DCM, concentration about 2–5%
- Addition of the corresponding amount of support (only a part should be coated, see Table 8.1) to the polymer solution, slight heating with stirring by hand and evaporation of the solvent (ventilator and exhauster)
- Careful crushing of clumps; quantitative transfer of the coated support from the beaker
- Filling the column (after the outlet is closed) with degassed nonsolvent mixture according to the interstitial volume (see Table 8.1). The support is added in two steps: (i) uncoated support (90–95%) and (ii) coated support on the top. The support must always be covered with liquid.
- Setting up the column (Fig. 8.1) and mixing vessel (Fig. 7.1)
- Filling the mixing vessel with degassed nonsolvent mixture

- Filling the syphon with degassed nonsolvent mixture, see p. 113. The syphon must be filled so that it is free of air.
- Filling the degasser and the storage vessel with solvent mixture. Refilling of the storage vessel may be necessary.
- Thermostatting of the degasser
- Adjustment of the temperature gradient. The desired temperature difference must be adjusted by variation of amperage (*only low tension must be used*!) and water flow. With the latter method it is better to adjust *after* the hot water outlet.
- Start of the elution by opening the foot valve of the column and regulation of the desired flow rate. Flow rate must be monitored and adjusted repeatedly.
- Collection of the sol fractions in suitable flasks or beakers depending on number and mass of the fractions. Generally, the eluate can be divided into as many fractions as you like.
- Elution of the column with a plenty of pure solvent or at some higher foot temperature after the last fraction (purging of support and column)
- Processing of the fractions by direct drying in a vacuum. In this case, empty but identically treated (washed, dried, weighed) beakers should also be weighed to eliminate adsorptions on the glass surfaces of the beakers.

**Evaluation**

- Determination of mass and MW or related measures ($[\eta]$, $\varphi^*$, $V_e$ from SEC) for each fraction
- Comparison of gradation and sharpness of fractions and starting polymer by SEC or turbidimetric titration (are inversions found?)
- Comparison of directly measured and calculated averages of MW or $[\eta]$ for the whole sample according to Eqs. (2.2) to (2.4) and (4.1).

Conclusions about limits, course (steepness), and parameters of the solvent/nonsolvent gradient and about the width of the temperature gradient should be deduced to optimize the fractionation (see Sects. 7.1.1 and 9.1).

## References


1. Baker CA, Williams RJP (1956) J Chem Soc [London]: 2352
2. Porter SR, Johnson JF (1967) In: Cantow MJR (ed) Polymer fractionation. Academic, New York, 95
3. Barrall EM, Johnson JF, Cooper AR (1977) In: Tung LH (ed) Fractionation of synthetic polymers. Marcel Dekker, New York, p 267
4. Cooper AR (1978) In: Epton R (ed) Chromatography of synthetic and biological polymers, vol 1. Ellis Horwood, Chicester, p 240
5. Glöckner G (1987) Polymer characterization by liquid chromatography. Elsevier, Amsterdam

6. Cantow HJ, Seifert E, Kuhn R (1966) Chemie-Ing-Technik 38: 1032; Cantow HJ, Probst J, Stojanow C (1968) Rev gen Chaoutchoucs et Plastiques 45: 1233; (1968) Kautschuk-Gummi-Kunststst 21: 609
7. Polaček J (1963) Coll Czech Chem Commun 28: 3011
8. Schulz GV, Deussen P, Scholz AGR (1964) Makromol Chem 69: 47
9. Glöckner G (1968) Faserforsch Textiltechn 19: 82

# 9 Trouble Shooting and Large-Scale Techniques in Column Fractionations

The fractionation efficiency of column-fractionation techniques (column extraction, Sect. 6.4; gradient elution, Sect. 7; Baker-Williams fractionation, Sect. 8) is partly influenced by the same parameters. Therefore, a joint discussion of these and some special parameters is advantageous. Besides, the same columns can be used and similar limitations hold [1].

## 9.1 Trouble Shooting and Fractionation Efficiency in Column Fractionations

Fractionation efficiency depends on the optimal choice of all adjustable parameters of the procedure [1]. The most common parameters of fractionation steps, possible problems, and consequences of these shortcomings are summarized in Table 9.1. The general result of all these is reduced fractionation efficiency and frequently the occurrence of inversions in the sequence of fractions, especially in the high-MW range (reverse-order fractionation, cf. Sect. 3.2). The efficiency of a fractionation can be judged by measuring the non-uniformity $U$ of the resulting fractions (SEC, subfractionation, turbidimetric titration, ultracentrifuging), by the calculated overall non-uniformity of the starting polymer based on fractions, and by the proportion of the inverted fractions (cf. Sect. 4.3.1).

**Load**

Let us now discuss Table 9.1. The *load* of the support is important in all column fractionations using a stationary gel phase. An overload can lead to clogging of the packing in all the techniques discussed. Partial clogging may cause inclusions of polymer species and increased flow resistance. Further clogging prevents any more fractionation because of column blocking.

If pre-fractionation does not occur in the loading procedure, fractionation efficiency is less in column extraction and gradient elution than in precipitation chromatography. This is because the effects of the latter, mentioned in Sect. 8.1, may partly compensate for the defect of insufficient coating.

**Flow Rates and Gradients**

A serious problem in precipitation chromatography, possibly caused by a too high flow rate, develops if the *gel film* does not form through the action of the

**Table 9.1.** Problems in column fractionation techniques (column extraction – CE, gradient elution – GEF, precipitation chromatography – PC)

| Step/parameter | Problem | Result[a] | Remarks[b] |
|---|---|---|---|
| Loading | – overload<br>– no prefractionation | – clogging inclusions | CE, GEF, PC<br>CE, GEF, (PC) |
| Sol–gel equilibrium | – no stationary gel phase (gel film) | – gel droplets<br>– rinsing out of small gel particles, turbid eluate, no chromatographic separation | PC<br>PC |
| Cross section | – too large | – sagging of the temperature profile | PC |
| Temperature gradient | – too steep | – sagging of the temperature profile | PC |
| Solvent/nonsolvent gradient | – too steep | – unsatisfactory sol/gel equilibrium | GEF, PC |
| Flow rate | – too high | – turbulent flow unsatisfactory sol/gel equilibrium<br>– sagging of the temperature profile | GEF, PC, (CE)<br>PC |
| | – too slow | – remixing of separated fractions | PC, (GEF, CE) |
| Degassing (unsatisfying) | – channelling<br>– gas accumulation in mixing vessel | – trouble with the solvent/ nonsolvent gradient | GEF, PC, (CE)<br>GEF, PC |

[a] General result of all the problems is a reduced fractionation efficiency and the common occurrence of inversions (reverse-order fractionation, cf. Sect. 3.2).
[b] Parentheses – has only a small influence.

temperature gradient. In such cases, if gel droplets or small movable gel particles result, the fractionation will be poor and the collected eluate turbid.

A very wide *cross section* of the fractionation column is a strong influence on precipitation chromatography, as discussed in Sect. 8.2. The same effect comes from a *temperature gradient* which is too steep (i.e., large values of $\Delta T$ per cm column length).

*Solvent/nonsolvent gradients* which are too steep hinder the full adjustment of the sol-gel equilibrium, especially in the high-MW range.

*Flow rates* too high or too low influence the stepwise column extraction less than the other procedures. Baker-Williams fractionation is particularly affected by troubles in the flow rate (cf. Sect. 8.3.2).

**Degassing**

Deficient *degassing* has consequences for the fractionation efficiency due to gas bubbles and channelling in the column bed because the action of gradients is disturbed. Gas accumulation in the mixing chamber diminishes the effective $V_m$ (Eq. (7.1)) and, thus, alters the course of the solvent/nonsolvent gradient.

Summarizing the discussion of Table 9.1, it must be stated that optimal relations of load, cross section, flow rate, solvent/nonsolvent gradient and temperature gradient can only be found experimentally.

The question arises which column technique possesses the highest *fractionation efficiency*. A general answer is impossible. Results and statements from the literature are contradictory. The efficiency depends chiefly upon optimal relations of all parameters discussed above. A critical analysis is given by Glöckner [2]. Two rough rules can be derived:

- If really optimal conditions can be set, precipitation chromatography yields a higher fractionation efficiency.
- Gradient methods can reach comparable fractionation results in a distinctly shorter time than stepwise column fractionation, but, the latter is less susceptible to disturbances and can be stopped at any time.

## 9.2 Large-Scale Column Techniques

Let us discuss the problem of large-scale fractionation for column extraction, gradient elution and precipitation chromatography, all of which use columns with a stationary gel phase. A review of the pioneering work in preparative-scale column fractionation has been given by Schneider [3].

Limitations of the mass of polymer fractionated arise from two parameters:

- diameter of the fractionation column and
- load of the support.

**Mass of Polymer**

The most common load is about 0.2 mg polymer per $cm^2$ support surface. In some large-scale fractionations (see Table 9.2) considerably higher loads are used (a few mgs/$cm^2$). When a temperature gradient is not used, the support may be coated totally. In such a case one is able to fractionate about 10 g polymer without problems in a column with a length of 100 cm and a diameter of 2.5 cm: The equivalent support volume of about 490 $cm^3$ (appr. 900 g glass beads with 0.4 mm diameter) possesses a surface area of ca. 5 $m^2$ leading to a load of about 0.22 mg/$cm^2$ which is a common value in column fractionation. With a load of 1.7 mg/$cm^2$ as used by Cantow et al. [4–6] in precipitation chromatography (the support was only partly coated!), about 75 g polymer could be fractionated without a temperature gradient.

The most important prerequisite for a successful large-scale fractionation is the fractionating deposition of polymer on the support.

**Table 9.2.** Large-scale column fractionations (polymer mass >10 g)

| Author(s) (year) [Ref.] | Polymer | Mass (g) | Column diam. (cm) | Column length (cm) | Method[a] | Remarks |
|---|---|---|---|---|---|---|
| Henry (1959) [9] | PE | 50 | 7 | 120 | CE | Support: Celite |
| van Schooten et al. (1961) [10] | PP | 30* | | | GEF | *extract from a 1000 g sample |
| Cantow et al. (1961) [4] | PIB | 34 | 6*2.5 | 100 | PC | six parallel columns |
| Cantow et al. (1963) [5] | PIB<br>PVAC | 50<br>40 | 6*2.5 | 100 | PC | six parallel columns |
| Cantow et al. (1964) [6] | Polyester | 90 | 6*2.5 | 100 | PC | six parallel columns |
| Myers and Dagon (1964) [11] | Poly (hydroxy ether) | 30 | 7.6 | 213 | GEF | |
| Kenyon et al. (1965) [12] | PE | 100<br>500 | 5<br>10 | 600<br>600 | CE | Support: Celite |
| Slonaker et al. (1966) [7] | PE<br>PP<br>Polyether | 200<br>100<br>800 | 15 | 487 | PC | 220 l solvent + nonsolvent |
| Hendersen and Hulme (1967) [8] | PS<br>PBD<br>Butyl rubber | 43<br>44<br>45 | 10 | 100 | PC | 6 protruding baffles |
| Lovric et al. (1976) [13] | S/MMA | 12 | 2.6 | 120 | CE | Support: glass wool |

[a] CE – stepwise column extraction, GEF – gradient-elution fractionation, PC – precipitation chromatography.

**Column Dimensions**

The largest column diameter known from the literature is 15 cm (6 in.) used in precipitation chromatography [7]. The problems arising from use of temperature gradients are discussed in Sect. 8.2 (temperature profile). Cantow et al. [4–6] overcame these difficulties by using six parallel columns inside a common temperature jacket, each column with a diameter of 2.5 cm. Henderson and Hulme [8] used six metal groins protruding from the wall into the column bed to guarantee an equal temperature profile in a column with 10 cm diameter.

A survey of large-scale column fractionations with more than 10 g polymer is given in Table 9.2.

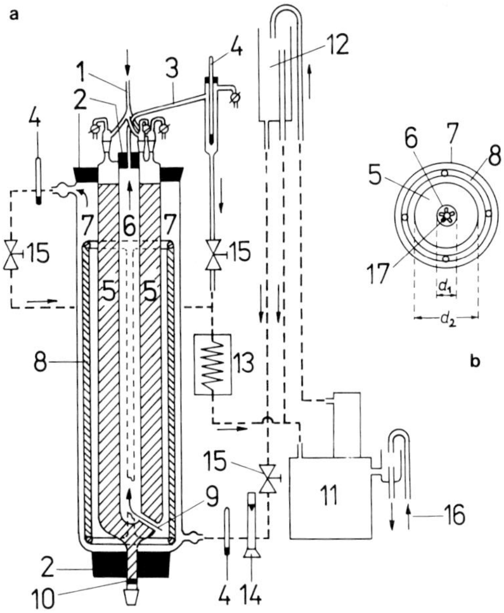


**Fig. 9.1a,b. a** Preparative all-glass column [14] and water circuit for precipitation chromatography (schematically). Heating device, heating wires, and inner support for wires are omitted. **b** View to the column from above (the rubber stoppers are removed). *1* – inlet of solvent/nonsolvent mixture, *2* – rubber stoppers, *3* – water outlet of the inner jacket, *4* – thermometers, *5* – fractionation bed, *6* – inner jacket, *7* – outer jacket, *8* – outer support for heating wires, *9* – water flow into the inner jacket, *10* – glass frit *G1*, *11* – recirculating thermostat, *12* – overflow vessel, *13* – cooling device, *14* – flow meter, *15* – valves, *16* – outer water circuit, *17* – five-armed inner support

A column with ring-shaped cross section made completely of glass [14] and shown in Fig. 9.1 has been successfully used in our laboratory. Parameters of this column are as follows: length 100 cm, diameter $d_1 = 2.5$ cm and $d_2 = 7.0$ cm, 3360 $cm^3$ support, 5 inner and 32 outer heating wires (78 Ohm/m), usually 2 l mixing chamber. The ring-shaped cylinder is embedded between an inner and an outer water jacket for thermostatting or for the generation of the temperature gradient. The latter problem is solved in the same way as in the smaller (analytic) column shown in Fig. 8.1 by use of heating wires and flowing water. A proposal for the design of the water circuit is also shown in Fig. 9.1. Temperature gradients can be adopted by coordinated regulation of water flows (valves 15) and electrical-power output (only low tension must be used!). 5 g polystyrene (about 0.2 $mg/cm^2$) was successfully fractionated according to Baker and Williams using flow rates of 900 to 1100 ml/h (26.8 to 32.7 $ml/h \cdot cm^2$).

This column was used to fractionate about 75 g polymer in a gradient elution without temperature gradient and with totally coated support under the same conditions as discussed above for the smaller column (about 0.2 $mg/cm^2$).

Large-scale column fractionations usually require large mixing vessels (this can be calculated using Eq. (7.1)) and therefore large amounts of nonsolvent mixtures "A". The same solvent/nonsolvent gradient can be generated stepwise using a smaller mixing chamber. This results in economies in the consumption of solvent and nonsolvent in comparison to a larger mixing vessel. A detailed description of such a variant is given in the Appendix (A 9).

The polymer masses mentioned in this section can be fractionated into several fractions within one run. This is an advantage over continuous polymer fractionation (cf. Sect. 6.6.2) which yields only two fractions per run (in unlimited mass, however). Besides, the relations between polymer mass fractionated, time required for fractionation, and volumes of solvents and nonsolvents are relatively favourable in large-scale column techniques.

## References

1. Barrel EM, Johnson JF, Cooper AR (1977) In: Tung LH (ed) Fractionation of synthetic polymers. Marcel Dekker, New York, p 267
2. Glöckner G (1987) Polymer characterization by liquid chromatography. Elsevier, Amsterdam
3. Schneider NS (1965) J Polym Sci C 8: 179
4. Cantow MJR, Porter RS, Johnson JF (1961) Nature [London] 192: 752
5. Cantow MJR, Porter RS, Johnson JF (1963) J Polym Sci C 1: 187
6. Cantow MJR, Porter RS, Johnson JF (1964) J Appl Polym Sci 8: 2963
7. Slonaker DF, Combs RL, Guillet JE, Coover HW (1966) J Polym Sci A-2, Polym Phys 4: 523
8. Henderson JF, Hulme JM (1967) J Appl Polym Sci 11: 2349
9. Henry PM (1959) J Polym Sci 36: 3
10. van Schooten J, van Hoorn H, Boerma J (1961) Polymer 2: 161
11. Myers GE, Dagon JR (1964) J Polym Sci A 2: 2631
12. Kenyon AS, Salyer IO, Kurz JE, Brown DR (1965) J Polym Sci C 8: 205
13. Lovric L, Grubisic-Gallot Z, Kunst B (1976) Europ Polym J 12: 189
14. Glöckner G (unpublished results)

# 10 Temperature Rising Elution Fractionation

## 10.1 Principles and Limitations of Application

Temperature rising elution fractionation (TREF) is a special kind of column-fractionation technique developed for semi-crystalline polymers, especially for poly(ethylene), poly(propylene) and their copolymers. TREF is used to characterize the distribution of short-chain branching (SCB) via a fractionation technique. Informative reviews on the method have been published recently [1, 2].

A semi-crystalline polymer can be fractionated with respect to MW above the melting point of the crystallites (cf. Sects. 3.1 and 4.1). To that end, gradient-elution fractionations at higher temperatures are possible [3, 4].

Crystallization and re-dissolution of crystallized polymers depend on MW and can also be used for fractionation on a large scale according to MW (crystallization-dissolution fractionation [5, 6]). However, this method is effective only in the low-MW range up to an order of magnitude of $10^4$ and for linear structures. SCB superimposes this procedure.

TREF is a column method with temporally increasing elution temperature which, on the face of it, resembles gradient-elution procedures with temperature program (cf. Sects. 7.1 and 7.2). TREF works, however, usually at higher temperatures and pursues another end, i.e., fractionation according to SCB.

Fractional crystallization and melting based on the theory of melting-point depression (see [1]) leads to decreasing elution temperatures with growing SCB and, consequently, to fractionation according to crystallizability due to differences in SCB. Influence of the MW is negligible.

The polymer to be fractionated must be fractionally deposited on an inert support by programmed cooling slow enough for crystallization of the different branched species in layers one upon the other. This step favours the effective separation by elution with increasing temperatures, which may be carried out stepwise or continuously.

The following features of TREF have been observed [1]:

- Growing SCB or increasing comonomer content results in an almost linear decrease of the melting and elution temperature which is the basis of TREF.
- The packing material has almost no influence.
- Cocrystallization of species with different degree of branching was not ascertained.

The development of the TREF method was influenced by gradient-elution fractionation. The design of TREF is similar to this method. *Analytical TREF* is largely automated and usually calibrated to determine SCB (number of methyl groups per 1000 C) from elution temperatures. This calibration results in a straight line for all samples with the same type of branching [1]. Of course, SCB can be determined also directly, e.g., by IR spectroscopy. The sample size ranges from 2 to 200 mg.

*Preparative TREF* is performed to isolate fractions for further investigations. It is usually based on results of analytical TREF investigations and in general a scaling-up procedure of the analytical variant. The masses of polymer fractionated by preparative TREF lie usually in between 1 and 10 g.

Owing to the different aims of TREF and the other column fractionations, a direct comparison is difficult. Nevertheless, the following advantages and disadvantages can be named.

**Advantages.**
- Unique method for separation according to SCB
- Almost linear relationship of elution temperature and degree of branching
- Easy automation and convenient handling
- Fractionation of very small samples and likewise scaling-up of the method
- Possible extension to cross fractionation (see Sect. 12).

**Disadvantages.**
- High-temperature handling
- Application of antioxidants or protecting gas.

TREF is important for fractionation of different types of poly(ethylene), poly(propylene), copolymers (EVA) and polyolefine blends.

## 10.2 Equipment and Materials

Simple TREF systems are built up very similar to gradient-elution fractionation (cf. Sect. 7.3). An example is shown in Fig. 10.1 containing all essential devices.

**Separate Parts of TREF Sytem**

**Solvent reservoir and degasser.** If the solvent reservoir is heated to the operating temperature, a separate degasser can be omitted. Otherwise, a degasser is necessary.

**Pump.** If the solvent does not flow by gravity (as in Fig. 10.1), different pumps, e.g., from SEC or LC, can be used.

**Column.** The column, filled with an inert support, can be manufactured from glass or (usually) steel. Glass columns are similar to those described in Sects. 6 to 8 (cf. Figs. 6.2, 7.1, 8.1, and 10.1). Metal columns having very different

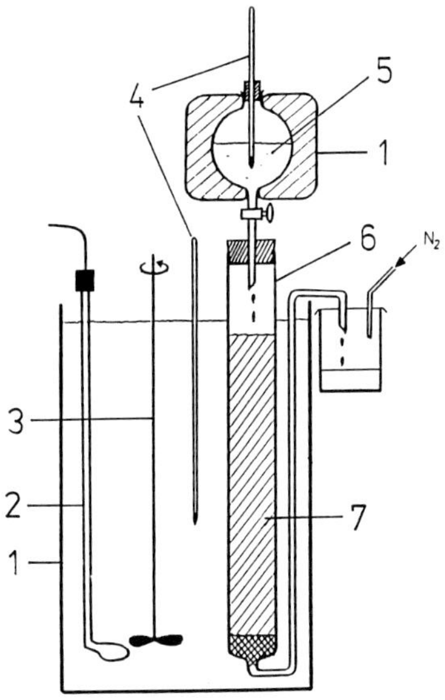


**Fig. 10.1.** Example of a simple TREF apparatus working by gravity. Column dimensions: $d = 7$ cm, $l = 38$ cm; support sea sand. *1* – thermostat, *2* – heater, *3* – stirrer, *4* – thermometer, *5* – solvent, *6* – column, *7* – support (Ref. [4], K. Shirayama et al., J. Polym. Sci. A 3 (1965), Copyright: John Wiley & Sons, Inc. Reprinted by permission of John Wiley & Sons, Inc.)

dimensions are mostly similar to columns used in LC or SEC. Column dimensions vary from 60 × 12 cm (*preparative* TREF) to 7 × 0.9 cm (*analytical* TREF). For survey, see, e.g., [1].

**Programmed heating device.** Heating and cooling of the column can be done stepwise or continuously (see Sect. 10.3). Different possibilities are used:

- Column jacket or bath filled with oil and connected with stirrer or recirculating pump. This variant is advantageous when thick columns are used.
- Ovens with programmed temperature control from SEC, LC or GC. This equipment is more convenient to handle, but, owing to the poorer heat transfer of air, problems arise when thicker columns are used.

**Fraction collector.** Fractions are usually collected only in preparative TREF.

**Detectors.** Refractive index (RI) and infrared (IR) detectors are used whereby IR detectors offer the advantage of a better baseline stability at high temperature. Detectors must operate at the same temperature as the column or somewhat higher. Connecting tubes must also be heated.

TREF systems can be coupled directly with SEC or continuous viscometry [7].

## Materials

**Support.** The following support materials are used [1]: Sea sand, glass beads, Chromosorb P, silanated silica gel, firebrick and steel shot. The amount depends on the volumes of the columns used.

**Solvent.** The high-boiling solvents xylene, trichlorobenzene, o-dichlorobenzene, kerosene, and $\alpha$-chloronaphthalene are used.

Concerning the volume of solvents necessary for one run, detailed data are not given. An estimation leads to about 1 to 5 l depending on dimensions of the column, elution mode and sample size. Additionally, a smaller solvent volume is required for deposition of the polymer on the support (see Sect. 10.3.1).

The application of **antioxidants** or **a protecting gas** (nitrogen) is recommended to protect the hot polymer solution against oxidation and degradation.

## 10.3 Specified Steps of Implementation

### 10.3.1 Preparatory Investigations

**Choice of Solvent**

The following aspects must be considered:

- Range of elution temperature
- Further measurements with the eluted fractions (e.g., SEC, viscometry)
- Kind of detection, i.e., sufficiently dissimilar refractive indices for RI detection or no mutual disturbances of the IR spectra of polymer and solvent
- Nontoxicity.

**Choice of Temperature Program**

Limits of the temperature range can be determined by solubility and precipitation tests at different temperatures.

The determined temperature difference can be crossed stepwise or continuously. The best procedure must be tested experimentally in analytical TREF runs. The following temperature programs were successfully used [1]:

**Stepwise elution** is mostly used in preparative TREF. Temperature rising to the desired level is followed by isothermal setting for different periods of time dependent on temperature step and column dimensions. This period of isothermal equilibration should be the longer the higher the temperature jump and the thicker the column are. Periods lasting from 10 minutes up to nearly one hour have been observed as necessary for equilibration.

The elution in each step can be continuously or can be interrupted for equilibration (stop-flow elution). The last version is recommended. Controlling of the time necessary for complete elution in the respective step can be made by addition of nonsolvent in excess to the eluate (turbidity) or by comparison of SEC runs after different TREF elution times.

Number and height of temperature steps must be chosen according to desired number and mass of fractions. When one wishes fractions of equal masses, temperature differences might be unequal depending on the shape of the SCB distribution.

**Continuous elution** is a feature of analytical TREF although stepwise procedures are possible. The continuous elution by the solvent is combined with a continuous temperature rising (8 to 50 K/h [1]). This is possible only when small and thin columns are used.

### 10.3.2 Fractionation Steps

**Deposition of Polymer**

Coating the support can be performed outside as well as inside the column. The latter is more usual and convenient.

Generally, polymer solutions above the highest fractionation temperatures are injected into the heated column packed with the support. The following slow cooling process connected with the fractional deposition of the polymer is the most important step for an efficient TREF. The more differentiated the deposition of the polymer occurs the more complete is the fractionation with respect to SCB. Different modes and rates of cooling are used:

- Natural cooling which is a relative fast process
- Programmed cooling with constant rate
- Programmed cooling with decreasing rate.

Constant cooling rates of 1 to 2 K/h seem to be optimal. Attempts to improve the fractionation efficiency have been made with about 30 K/h followed by descending rates up to 2 K/h. The periods of time used for the cooling procedure range from a few hours to a few days.

The fractionation efficiency is higher the smaller the ratio of coating polymer to support (i.e., the thinner the coating) is. Inspection of the literature data leads to loads in the order of magnitude of $10^{-2}$ to $10^{-4}$ g/cm$^3$ (polymer/bulk volume of the support) where most frequently $10^{-3}$ g/cm$^3$ has been used, independent of sample size, mode of TREF (analytical or preparative) and kind of support.

It is advantageous to use the support completely for loading. Therefore, the volume of the polymer solution injected into the packed column should correspond to the liquid volume inside the column.

**Degassing**

Degassing must be done above the highest fractionation temperature.

**Heating and Elution**

Possible heating regimes are discussed above (preparatory investigations). If the temperature rise is too steep, equilibration in the deposited polymer is hampered and the fractionation efficiency is reduced. Elution rates range from 1 to about 5 ml/min in different modes (see above).

**Collection and Processing of Fractions**

The eluates of analytical TREF runs may be analysed on-line or off-line by SEC.

Fractions from preparative TREF can be used in solution for further analyses, but usually, the polymer is precipitated by cooling and often additionally by a nonsolvent. In the latter case, low-boiling nonsolvents are recommended to facilitate drying of the fractions.

**Parameters Reducing TREF Efficiency**

- Insufficient fractional deposition of the polymer and overloading of the support
- Insufficient thermal equilibrium inside the column caused by diameters which are too large, temperature changes which are too steep, and poor heat transfer
- Channelling, especially in thick columns.

## 10.4 Example for a TREF of Poly(ethylene) (LDPE) with Xylene

The example has been formulated so that the fractionation can be carried out on different scales. A compilation of some operating variables in relation to the dimensions of the column used is summarized in Table 10.1. For other dimensions, the values can be estimated analogously.

**Materials** The amounts of polymer and solvent can be inferred from Table 10.1.

**Polymer.** LDPE with a density of about 0.92 $g/cm^3$

**Table 10.1.** Operating variables of TREF (LDPE/xylene)

| Column dimensions: | | | |
|---|---|---|---|
| $l$ (cm) | 20 | 50 | 50 |
| $d$ (cm) | 2.5 | 2.5 | 7.0 |
| Support (bulk volume) ($cm^3$) | 100 | 250 | 1900 |
| Polymer[a] (g) | 0.5 | 1.25 | 9.5 |
| Solution volume[b] (ml) | 26 | 65 | 495 |
| Flow rate (ml/min) | 2 | 4 | 10 |
| Eluate volume per step[c] (ml) | 60 | 150 | 1000 |
| Overall eluate volume (ml) | 600 | 1500 | 10000 |
| Average mass of fractions (g) | 0.06 | 0.15 | 1.1 |
| Period of fractionation[d] (h) | 10 | 12 | 24 |

[a] Values are related to a load of about $5 \times 10^{-3}$ $g/cm^3$ (polymer/bulk volume of the support).

[b] Interstitial volume for deposition of the polymer. The load of $5 \times 10^{-3}$ $g/cm^3$ can be reached with a polymer concentration of ca. 15 g/l. The given volumes are calculated values; the exact liquid volume inside the packed column should be determined experimentally.

[c] Estimated volume of the eluate in each tempearture step after 30 minutes for equilibration (stop-flow elution).

[d] Without preparatory steps.

**Solvent.** Xylene (addition of 0.1 or 0.2% antioxidant is recommended)

**Nonsolvent.** Methanol or acetone (analytical grade, for precipitation), a volume about ten times the volume of the eluted fraction (partly reduced by evaporation)

**Support.** Sea sand or glass beads (diameter about 0.3 mm); the amount can be taken from Table 10.1.

**Equipment**

- Column similar to Figs. 6.2 or 10.1 equipped with a jacket and recirculating pump. Placing in an oil bath is also possible. Any glass tube with a length between 20 and 50 cm and a diameter in the range from 2.5 to 7 cm can be used (see Table 10.1).
- Storage vessel and degasser: The storage vessel may be heated or connected with a degasser. When a column is used which is considerably longer than necessary for TREF, the unpacked volume above the fractionation zone can serve as a preheater and degasser of the solvent.
- Devices for collection and processing of the fractions: The outlet tubes and valves must be heated to prevent precipitation of the polymer inside the apparatus. Beakers and similar vessels for collection, precipitation and drying of about 10 fractions are required.
- Heating and monitoring device for a temperature range from room temperature to about 120 °C.
- Glassware for dissolution of the polymer and for filling the column, for precipitation, and further processing of the fractions
- Stands, clamps, sockets etc.

**Time Required**

- Dissolution of polymer — 2–3 h
- Deposition of polymer — about 40 h
- Thermostatting — 0.5 h
- Elution at constant temperature (depending on column dimensions and mass of polymer) — 0.5–1.5 h

Thermostatting and elution time are valid for each fraction. Time necessary for processing of the fractions is not taken into account.

**Preparatory Investigations**

- Determination of the temperature range: According to Sect. 10.3, temperatures for complete solubility and total insolubility (by precipitation tests after dissolution tests at increasing temperatures) must be determined. They might lie for the sample fractionated at about 100 °C and 50 °C, respectively.
- Dissolution of the polymer: The polymer necessary for coating the required support volume (see Table 10.1) with a load of about $5 \times 10^{-3}$ g/cm$^3$ is dissolved in the equivalent solvent volume (see Table 10.1) at approx. 120 °C (c = 15 g/l).

**Fractionation** Experimental conditions can be chosen from Table 10.1.

- Deposition of the polymer: Two variants are possible.
- The heated and empty column is filled with that volume of heated polymer solution (with a concentration of about 15 g/l) which is necessary for total coating of the support bed inside the column (see Table 10.1). This volume should be determined in advance. The heated support is then loaded into the column reaching finally an equal surface of support and solution.
- The packed column filled with the solvent is heated to about 120 °C. After this, the solvent inside the packed support is displaced by the appropriate volume of heated solution (by gravity or by means of a pump).

Now, after both variants, the solution inside the packed column is cooled at a rate of about 2 K/h near to room temperature.

- After connecting to the degasser and the storage vessel, the temperature is raised to 50 °C; elution is started after about 30 minutes: Flow rate ca. 2 ml/min for narrow columns and 5 to 10 ml/min for wide ones. The elution is interrupted when no turbidity of the eluate occurs after cooling and addition of excess nonsolvent. Estimated values of these solvent volumes are given in Table 10.1.
- After the elution of the first fraction, the apparatus is heated up to 60 °C, equilibrated (30 min) and the column afterwards eluted at 60 °C.
- In this way (stop-flow elution), all other temperature steps (fractions) are performed at 70, 75, 80, 85, 90, 100 °C. Hence, eight fractions of almost equal mass (depending on SCB distribution) should be obtained.
- Finally, the column should be eluted briefly at a temperature of about 120 °C.
- The fractions can be used for further investigations after precipitation and drying. Depending on the dimensions of the column, fractions of about 50 mg to roughly 1 g can be obtained.

**Evaluation**

- Determination of mass and degree of SCB for each fraction
- Mass balance of TREF and of the fractions mutually
- Determination of MW or MWD of the fractions.

## References

1. Wild L (1990) Adv Polym Sci 98: 1
2. Glöckner G (1990) J Appl Polym Sci, Appl Polym Symp 45: 1
3. Wijga PW, van Schooten J, Boerma J (1960) Makromol Chem 36: 115
4. Shirayama K, Okada T, Kita S (1965) J Polym Sci A 3: 907
5. Pennings AJ (1967) J Polym Sci C 16: 1799
6. Koningsveld R, Kleintjens LA, Geerissen H, Schützeichel P, Wolf BA (1989) In: Booth C, Price C (eds) Comprehensive polymer science, vol 1. Pergamon, Oxford, p 293
7. Nakano S, Goto Y (1981) J Appl Polym Sci 26: 4217

# 11 Partition Fractionation

The distinguishing feature of partition fractionation is the distribution of the polymer species between two immiscible phases which must exist even without the polymer to be fractionated. This immiscibility of the liquid phases differentiates *partition between immiscible liquids* from coacervate extraction (cf. Sect. 6.5).

Partition of polymer species may also proceed when only one phase is liquid and the other a bonded polymer phase. This so-called *phase-distribution chromatography* differs from other column-fractionation procedures discussed in the foregoing sections because outer gradients to manipulate the solubility are absent. This method will be considered in Sect. 11.2 whereas liquid–liquid partitions will be discussed in Sect. 11.1.

A survey of different techniques of partition fractionation is given in Table 11.1.

## 11.1 Partition Between Immiscible Liquids

### 11.1.1 Principles and Limitations of Application

#### 11.1.1.1 General Considerations

The basis of this fractionation technique was laid down by Schulz [1, 2]. Systematic studies were performed by Almin [3], v. Tafel et al. [4–8], and Kuhn [9–12].

Equation (3.8) holds generally for the partition of polymer species with $P_i$ between two immiscible phases [2].

**Phase Relations**

In terms of the above outlined fractionation techniques, single and double primed quantities represent the sol phase (as a rule, the upper phase) and the gel phase (as a rule, the lower phase), respectively. Here, gel and sol phases in their usual meaning do not exist. The highest-MW components can be enriched generally in the upper or lower phase. In this section, all double primed quantities represent the phase where the highest-MW species are concentrated ("gel-analogous" phase) while all single primed values denote the "sol-analogous" phase with the lower-MW components. For fractionations with respect to CC, single and double primed quantities indicate the upper

**Table 11.1.** Variants of partition fractionation

| Procedure | Partition between immiscible liquids | | | | | |
|---|---|---|---|---|---|---|
| | Simple partition (one- and multiple-step technique) | | | Counter-current partition (Craig) | Counter-current chromatography | Phase-distribution chromatography |
| Temperature: | | | | | | |
| constant | X | | X | X | X | X |
| variable | | X[b] | | | | |
| Solvent composition[a]: | | | | | | |
| constant | X | X | | X | X | X |
| variable | | | X | | | |
| Support: | | | | | | X[c] |
| Mode of procedure: | | | | | | |
| stepwise | X | X | X | | | |
| counter-current | | | | X | | |
| continuous (column) | | | | | X | X |
| chromatography | | | | | X | X |
| preparative | X | X | X | X | X | |
| analytical | X | X | X | X | X | X |
| Direction[d]: | | | | | | |
| molecular weight | X | X | X | X | X | X |
| chemical composition | | X[b] | X | | | |

[a] Summative composition of all conjugated phases.
[b] A special variant is the fractionation with demixing solvents caused by decreasing tempearture.
[c] Support is coated with polymeric bonded phase.
[d] Fractionation according solely to MW is mainly possible only for polymers with constant composition.

and lower phase, respectively, $K_i$ is defined as in Eq. (3.8). Consequently, Eqs. (3.10) and (3.11) do not change for the "gel-analogous" phase and for the "sol-analogous" phase, respectively.

**Distribution Coefficient**

The *distribution coefficient* $K_i$ grows with increasing values of $k$ and $P_i$ resulting in enrichment of the high-MW components in the "gel-analogous" phase. Simultaneously, the partition between the phases is usually strongly asymmetrical, i.e., $K_i$ values are very high ($\gg 1$) or very low ($\ll 1$). The asymmetry of the partition increases with ascending MW.

"Gel-analogous" phase and quantity $k$ must be considered a little bit more in detail to understand factors ruling $K_i$. In partition fractionation, the polymer is fully soluble in the "gel-analogous" phase, i.e., this phase has a higher dissolution power than the "sol-analogous" phase. The higher the difference in solvent strength of the conjugated phases the more asymmetrical is the distribution of the polymer species. To reduce this effect, the dissolution power can be either lowered in the "gel-analogous" phase or enhanced in the "sol-analogous" phase, e.g., by temperature variation or by an additional component. This component alters the value of $k$ which represents the difference in solvent strength (expressed by $\Delta\chi$) of the immiscible phases for the given polymer:

$$k \propto \Delta\chi = \chi' - \chi'' > 0 \tag{11.1}$$

A decrease in $k$ which reduces the asymmetry of the partition, can be achieved by an increase of $\chi''$ or a decrease of $\chi'$. In a binary liquid mixture (at constant pressure), the dissolution power can be influenced by temperature variation – this alters the composition of the conjugated phases and consequently the values of $\chi'$, $\chi''$ and $k$. The same can be achieved at constant temperature by addition of a third component and its special distribution between the "gel- and sol-analogous" phases. $\Delta\chi = 0$, $k = 0$, and $K_i = 1$ are valid when $\chi' = \chi''$.

Increasing the amount of an additional component can even lead to an inversion of the phases, i.e., highest-MW species are accumulated in the previous "sol-analogous" phase which now becomes the "gel-analogous" phase and vice versa. Then, $\chi' - \chi'' < 0$ holds good (in original denotation of Eq. (11.1)) and $K_i$ becomes formally smaller than one.

Generally, the compositions and volume ratio of the conjugated phases are dependent (at constant pressure) upon temperature and composition of the solvent mixture. (In binary mixtures, compositions of the phases solely depend on temperature; total composition of the system influences the volume ratio.) The more dissimilar the compositions of conjugated phases the higher $k$ is and the more asymmetrical is the partition of the polymer between the phases. The following conclusions can be drawn:

- The dependence of $K_i$ upon MW opens the possibility of fractionation according to MW by partition (cf. Eq. (3.8)). Homopolymers and chemically homogeneous copolymers (which are to be fractionated with respect to MW) should be distributed between similar phases (near to the critical point, see next section) with values of $K_i$ close to one. The low fractionation efficiency caused by such $K_i$ values must be compensated by a multiple counter-current partition procedure [3–8].
- CC of the polymer strongly influences $k$. Therefore, fractionation (separation)of polymer blends [9–12] can be performed according to CC with only a small influence of MW.

- In the fractionation of chemically heterogeneous copolymers [4, 5, 9–15], Eq. (4.7), originally derived for sol-gel fractionation, also works generally in partition between immiscible liquids [14]. However, CC usually dominates in partition fractionation. MW may influence the fractionation only in the vicinity of the critical point [14, 15].
- $K_i$ is almost independent of concentration in diluted solutions [4].

#### 11.1.1.2 Selection of Solvent Mixtures

The following common *criteria of solvent selection* can be used [3, 4]:

- Immiscibility of the chosen solvents (at least below a certain temperature)
- Sufficient difference in densities of the solvents
- Sufficient dissolution capacity for the polymer (i.e., relative great portions of a good polymer solvent in the phases) to render high polymer concentrations
- Enrichment of at least one species (e.g., highest-MW component of a homopolymer or AN-rich portion of an AN-copolymer) in each phase
- Distribution coefficients $K_i$ close to unity (5–0.2, average for all polymer species) for counter-current distribution (see below)
- Accumulation of the parent homopolymers of copolymer constituents in opposite phases [10, 11].

The *search of suitable systems* may be implemented by means of Gibbs phase diagrams.

**Solvent Systems**

**Two components** [13, 14]. The system must exhibit an upper critical solution temperature (UCST, see Glossary), i.e., constituents are miscible above UCST and show a miscibility gap below. One component must be a good polymer solvent while the other should be a poorer solvent.

The one-phase region separates upon cooling into two phases of different composition. The phases should have different but sufficient dissolution capacity for the polymer species. Differences in composition, density, and dissolution capacity ascend with decreasing temperature. Simultaneously, distribution coefficients withdraw from unity.

An example of phase diagram is given later in Fig. 11.5 which will be discussed in Sect. 11.1.4.1.

Usually, homopolymers cannot be solubilized in two immiscible solvents (one of the rare exceptions is poly(oxyethylene)/water/chloroform [1]). Therefore, such systems are used chiefly for copolymers or polymer blends [9–15]. The best solvent pair for copolymers consists of solvents which dissolve just one of the respective parent homopolymers. Solubility parameters (cf. Sect. 3.1) can be used to estimate miscibility of solvents and polymers [9, 11].

Fractionations using UCST behaviour are often called as "*fractionations with demixing solvents*" (see Sect. 11.1.4).

**Three components** [3–5]. Partition fractionations of homopolymers require use of three components. The main component (A) should be a solvent of medium polarity mixed with a nonpolar (B) and a more polar (C) liquid. B and C are mutually immiscible, but at least one of them should be soluble in A. Solubility of the polymer in B and C is not necessary. The system can be illustrated in a phase triangle showing a miscibility gap for three components. Figure 11.1 gives an idealized schematic example for two temperatures. The following conditions would be optimal, but can be achieved only approximately:

- Extended and symmetrical miscibility gap (i.e., the critical point is situated in the vertex of the binodal which is not the usual case) to ensure a high solubility capacity and to avoid associates or aggregates in the solutions
- Horizontal tie-lines have constant relative amounts of the main component (solvent) in the conjugated phases.

Thus, phases of equal size can be easily obtained. This is especially advantageous in counter-current distribution (see below). The distance of a certain system to the critical point is directed by overall composition and temperature

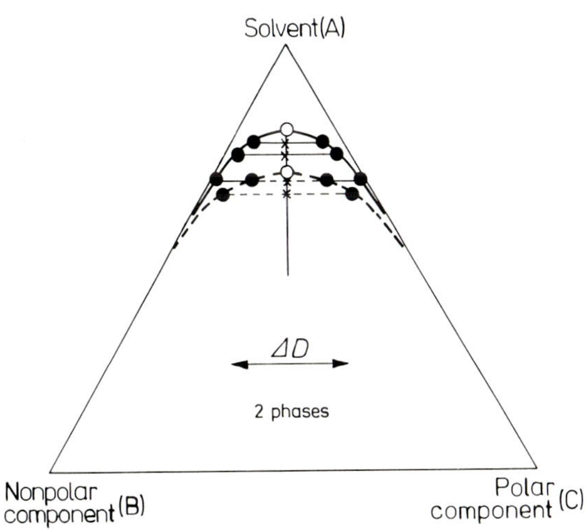


**Fig. 11.1.** Idealized phase diagram of a ternary liquid system with miscibility gaps at two temperatures $T_1$(——) < $T_2$(– –) (schematically). *Curves* are binodals with *critical points* (○). ● – compositions of the conjugated phases connected by *tie-lines*, × – overall composition of the system. The *vertical line* connects points of overall composition yielding conjugated phases of equal size. Differences in composition and density ($\Delta D$) of the phases are dependent on the distance of the system from critical point (influenced by overall composition and temperature – cf. the results for constant composition ( × ) and different temperatures on the binodals (–●– and – ● –)

(see Fig. 11.1). Increasing distance enlarges differences in composition and density of the phases, decreases dissolution capacity, and leads to more extreme distribution coefficients ($K_i \rightarrow 0$ or $\infty$). Because of last point, one works mostly with phase pairs near to the critical point, especially in countercurrent distribution.

**Four components.** These can be used when the polymer is still too asymmetrically distributed between a phase pair consisting of three components. In this case, the dissolution power of mainly one phase is changed by the additional component resulting in altered distribution coefficients. In extreme cases, even a phase change of polymer species ($K_i > 1 \rightleftharpoons K_i < 1$) may occur (cf. Sect. 11.1.1.1). An example for manipulation of $K_i$ is shown in Fig. 11.2 for the system PMMA/acetone/hexane/water/benzene [4, 5, 7].

It should be noted that the polymer may change the miscibility gap (see, for instance, [13]). Distribution coefficients cannot be read from the phase diagrams and must be measured in separate experiments (See Sect. 11.1.3.1).

It is worthwhile adding a few remarks about the *critical point* (see Glossary): At this point, the formerly separate phases have the same composition and density, and all polymer species should have a distribution coefficient $K_i = 1$ independent of MW. Deviations are said to be caused by end-group influences [3, 8]. MW-independent shifting of the point characterized by $K_i = 1$ is achieved by adding a small amount of a fourth component [4, 7]. However, this point is not identical with the critical point (see Fig. 11.2 – the conjugated phases do not have the same composition). This effect can be used to adjust optimal $K_i$ values in systems not too far from the critical point.

Details of the investigation of phase behaviour in the systems discussed above are given in Sects. 11.1.3.1 and 11.1.4.1.

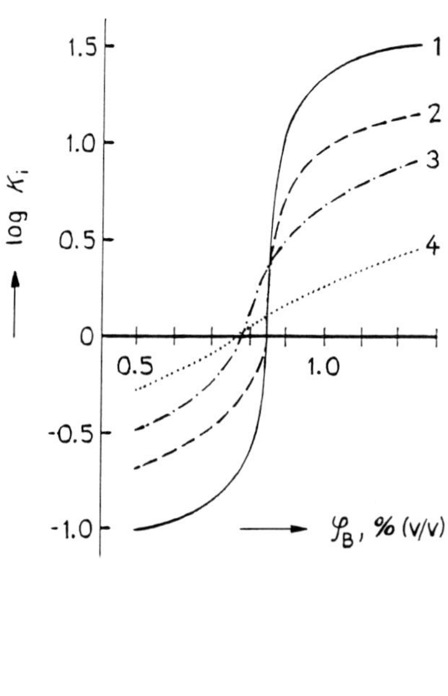


**Fig. 11.2.** Influence of benzene content (volume-%) on distribution coefficients of PMMA species of different MW in the system acetone/hexane/water. The composition of the conjugated phase pair is characterized by the difference in densities $\Delta D = 0.03$ g/cm$^3$. Enrichment of the polymer occurs in the (*lower*) acetone-rich water phase for $K_i < 1$ and in the (*upper*) acetone-poor hexane phase for $K_i > 1$. $K_i$ becomes unity for all species at about 0.8 volume-% benzene. MWs are: (*1*) 540000, (2)250000, (3) 130000, (*4*) 55000. (Adopted from Ref. [7] with permission of Hüthig & Wepf Verlag, Basel)

#### 11.1.1.3 Procedures

Partition fractionation by immiscible liquids can be performed as simple partition, counter-current partition, or as counter-current chromatography. Very many variants are known. Here, only a general description is possible. Every polymer/solvent-mixture system must be tested carefully. Table 11.2 gives a survey of applications of partition between immiscible liquids which may illustrate the experimental variety.

**Table 11.2.** Applications of liquid-liquid partition fractionation

| Author [Ref.] | Polymer[a] | Solvent system[a] | Remarks[b] |
|---|---|---|---|
| Schulz, Nordt [1] | POE | W/CF/B | MW, diff. comp., 16 g |
| Almin [3] | POE | EtOH/W/TCE | MWD, const. comp., |
| | | CF/W/TCE | CCP, 0.5–5 g |
| Hamann et al. [18] | PS + S/FA[c] | DMF/NM/ | Separation of species, |
| | PS + S/MA[c] | DHN/B | CCP |
| Case [16] | POE | W/CF/B | MW, const. comp., |
| | POP | W/Hx | CCP, up to 240 g |
| | POAlk | MeOH/W/CH | polymer |
| | | MeOH/W/CH/T | MW, diff. comp. |
| Fritzsche [17] | PAN | DMF/DHN | MWD, TV, 1.5 g |
| V. Tavel et al. | PMMA | Ac/Hx/W/B | MWD, CC, |
| [4–8] | PEMA | Ac/Hx/W/AA | CCP, const. comp., |
| | MMA/EMA[c] | Ac/Hx/W/AA | up to 4 g polymer |
| Kuhn [9–11] | binary | DMF/MCH, | |
| | mixtures | DMF/CH | const. comp., |
| | | (+3th comp.) | |
| | EVA/PS[d] | DMF/MCH | TV, CC |
| | LDPE/PS[d] | DMF/MCH | |
| | Additives | | Separation, TV |
| | Oligomers | | MW, separation, TV |
| | PS | DMF/CH | MW, TV |
| | S/BD[e] | DMF/Hp | |
| | | DMF/MCH | CC, TV, const. comp. |
| | | DMF/CH | |
| | CPE | DMF/MCH | CC, TV |
| | PDMS/PS[d] | DMF/MCH | CC, TV |
| Stejskal et al. [13] | PMMA/PDMS[d] | DMSO/TeCE | CC, TV, 4 g |
| Podešva et al. [14] | S/MEMA[c] | DMSO/TeCE | CC, TV |
| Podešva et al. [15] | S/HIP[e] | DMF/MCH | CC, TV |

[a] See abbreviations.
[b] CC – chemical composition; CCP – counter-current partition; TV – temperature variation; diff. comp., const. comp. – different or constant composition of the solvent system in partition steps, respectively.
[c] Binary statistical copolymers.
[d] Graft copolymers.
[e] Diblock copolymers.

**Simple partition.** Only a few partition steps are carried out. This procedure is chiefly used for fractionation and separation of different heterogeneous copolymers or polymer mixtures by demixing solvents [9–15]. Fractionation of homopolymers and copolymers with constant CC requires differences in the distribution coefficients of all species of $\Delta K_i > 10$ but, the efficiency is low.

In binary mixtures, extremely asymmetrical distributions often result, especially for homopolymers. For instance, poly(oxyethylene) (POE) in water/chloroform mixtures was almost completely found in the chloroform phase, i.e., the value of $k$ in Eqs. (3.8) and (10.1) is high. However, the partition of POE between the phases was improved by addition of the poor solvent benzene (diminution of $k$) which increasingly displaced the polymer species into the water phase the lower their MWs were [1].

Sometimes, different compositions, temperatures and phase volumes are used from step to step.

**Counter-current partition.** This procedure can be performed similarly to Craig partition under conditions of mostly equal phase volumes and constant phase composition near to the critical point. The overall distribution coefficient lies in the range 0.2–5 with $\Delta K_i < 10$ for all polymer species [3–8, 16]. This technique is commonly used in MW fractionation. However, fractionation with respect to CC is likewise possible when phase composition is successively altered [4–6].

**Counter-current chromatography.** In contrast to the last procedure, exchange and transport of the phases are continuous and simultaneous. This technique will not be discussed here; for a short review see [19].

**Use of auxiliary polymers.** This special variant suggested by Englert and Tompa [20], but used up till now only for biopolymers in aqueous systems [21], is based on incompatibility of solutions caused by *auxiliary polymers*. This technique, which can be performed as simple or counter-current procedure, will not be discussed in this book.

#### 11.1.1.4 Applications

The sequence of fractions in liquid–liquid partition fractionation can be different: increase as well as decrease of MW in simple ([16] and [1, 17]) or counter-current partition ([3] and [7, 16]), respectively. The course of a fractionation according to CC is dependent upon temperature differences between the fractionation steps, size of phases, and succession of their isolation. Figure 11.3 gives an example for a fractionation with demixing solvents and increasing fractionation temperatures for the consecutive fractions [13]. Other results, related to the sequence of fractions, can be expected by successive decrease of temperatures for individual fractions.

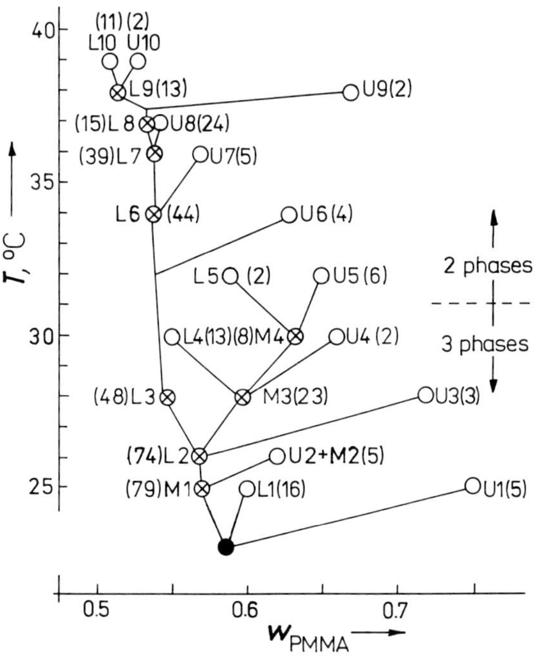


**Fig. 11.3.** Course of the partition fractionation of a PMMA-*g*-PDMS copolymer with demixing solvents (DMSO (30 volume-%) and tetrachloroethane (TeCE)) at increasing fractionation temperatures. ● – composition of the starting polymer in homogeneous solution at 60 °C; first fractionation step by cooling to 25 °C. Temperature was raised to 60 °C between each of the fractionation steps to homogenize the system. Solvent contents of the collected fractions were replaced by equivalent volumes of mixed solvents. *U* denotes the upper (DMSO-rich) phases, *L* the lower (TeCE-rich) and *M* the middle phases. ○ – collected phases, ⊗ – phases for further fractionation; percentages of the polymer in the phases (rounded and related to the total polymer mass) are given in parentheses. Above 31 °C and below this temperature, two and three liquid phases coexist, respectively. Data from Ref. [13]

All variants of liquid–liquid partition fractionation can be performed on analytical and preparative scales (cf. Table 11.2). This fractionation technique can be advantageously applied for polymers soluble in solvents of medium polarity. Success is doubtful with totally nonpolar (e.g., poly(ethylene)) or strongly polar polymers (e.g., polyelectrolytes).

Sometimes problems may arise: Emulsification of the system can occur; microseparation (supermolecular structures, copolymer micelles), thixotropy, and iridescence were observed in fractionation of a diblock copolymer [15], and combination of precipitation and partition mechanisms is likewise possible [14].

The following most important aspects of the fractionation of homopolymers, polymer blends, and copolymers can be summarised.

**Fractionation of homopolymers** (and chemically homogeneous copolymers) [1–8].

- Fractionation is usually performed at constant temperature.
- Ternary or quarternary solvent systems are mostly necessary.
- Use of constant or variable phase composition is possible.
- Multiple partition or advantageous counter-current technique is required.
- $K_i$ values should be close to unity, i.e., conjugated phases must have similar composition near to the critical point of the system.

**Separation of polymer blends** [9–12].
- Binary solvent systems are often sufficient.
- Demixing solvents and temperature variation are usually applied.
- Each species should be concentrated in one of the opposite phases.
- High $K_i$ values make one-step separations possible.
- Separation of homopolymers from block and graft copolymers may be possible.

**Fractionation of heterogeneous copolymers** [10, 13–15].
- Binary solvent systems are often sufficient.
- Demixing solvents and temperature variation are mostly applied.
- Parent homopolymers should be contained in opposite phases.
- Large differences in the composition of conjugated phases enables sharp one-step separation according to CC but no successive fractionation.
- Successive multi-step fractionation is possible in vicinity of the critical solution temperature or in the case of chemically similar monomeric units. In the former case, the influence of MW on the fractionation increases.

Let us finally summarize advantages and disadvantages of the partition-fractionation technique.

**Advantages.**
- Fractionation according to both MW and CC are possible and do not interfere when different fractionation conditions are used.
- Fractionation occurs completely in solution; adsorption and occlusion effects hardly take place; crystallizable polymers can be fractionated.
- Fractionation on a preparative scale and with simple experimental techniques is possible.
- Sharpness of fractionation grows with increasing MW because $k$ (cf. Eqs. (3.8) and (10.1)) remains almost constant during fractionation in contrast to other fractionation techniques.

**Disadvantages.**
- Expenditure to find suitable systems is high.
- Application is restricted to polymers of medium polarity.
- MW fractionation yielding sharp fractions is difficult.

- Method is time-consuming for systems near to the critical point (similar densities).
- Sometimes, emulsions hamper the separation of the phases.

The foregoing sections indicate that distinct differences exist in partition fractionations of homopolymers (and chemically homogeneous copolymers) according to MW and of polymer mixtures and chemically heterogeneous copolymers. Therefore, after description of equipment and materials necessary for both variants in Sect. 11.1.2, the experimental steps of fractionation according to MW or CC will be described in different sections (11.1.3 or 11.1.4, respectively).

### 11.1.2 Equipment and Materials

The apparatus required is similar to that used in coacervate extraction (cf. Sect. 6.5.1). For partition fractionation, some differences in equipment exist between preparatory tests, simple partition, and counter-current partition.

**Preparatory Investigations**

Standard glassware is necessary for measuring and dosing of liquid volumes, evaporation, drying, and weighing. Other equipment needed includes:

- Thermostat with glass window
- Ampoules or small separatory funnels
- Syringes for phase isolation
- Equipment for turbidimetric measurements (for some investigations, e.g., determination of binodals, see Fig. 11.1 and Sect. 11.1.3.1)
- For determination of phase composition: gas chromatograph, liquid chromatograph, densitometer, or differential refractometer (dependent on components).

**Simple Partition**

Only a small number of separatory funnels are usually required for simple partition procedures. Separatory vessels as shown in Figs. 6.1 (without frit) and 6.3 or cylindrical flasks can be used as an alternative to the funnels. The length-to-diameter ratio should be high in all the various kinds of vessels. Number and size depend on volume and number of the fractions.

Besides the standard glassware mentioned above one needs:

- Thermostat
- Devices for shaking, rocking, rolling (especially if emulsions can be formed) of the vessels or stirring inside the vessels
- Syringes or devices similar to Fig. 5.2 for phase isolation or transfer.

**Counter-Current Partition**

When counter-current partition is carried out by hand, the same equipment as in simple partition can be used together with a set of vessels suitable for intermixing, segregation and isolation of the phases. Phase pairs of similar density could require vessels suitable for centrifugation to accelerate the phase separation.

Many partition steps (> 10) and the corresponding number of vessels needs automation as in the procedure of Craig. A special version for conjugated phases of similar density is the partition centrifuge [7]. Such apparatus should work in a thermostated room. Devices for processing of fractions are additionally required for all variants.

**Materials** Table 11.3 shows a compilation of amount of materials, i.e., solvents, related to the mass of polymer to be fractionated, from literature. The overall polymer starting-concentration usually ranges from 5 to 20 g/l. Higher values were used by Case [16]. Starting polymer-concentration should decrease with increasing MW [4–8]. Since the volumes of the phase pairs vary between wide limits, concentrations are related to very different amounts of polymer. A comparable quantity is the relative volume (ml total solvent mixture per gram polymer) given in the last column of Table 11.3. Values show a distinct difference between counter-current partition according to Craig [3–8] and other partition procedures (see also Table 11.2).

**Table 11.3.** Relations between polymer masses and solvent volumes in partition fractionation

| Author [Ref.] | $c_0^a$ (g/l) | Mass of polymer[b] (g) | Starting phase pair[b] (l) | Sum of all phases[b] (l) | Relative volume (ml/g) |
|---|---|---|---|---|---|
| Schulz and Nordt [1] | ca. 8 | 16 (POE) | 2 W, 0.05 CF/B | 2 W, 0.3 CF/B | 144 |
| Almin [3] | 5 | 0.5 (POE) | 0.1 | 1–2 | 2000–4000 |
| Case [16] | 55 | 50 (POP) | 0.5 W, 0.4 Hx | 5 W, 4 Hx | 180 |
| | 88 | 75 (POE) | 0.7 W, 0.15 CF/B | 7 W, 1.5 CF/B | 113 |
| | 40 | 29 (POE) | 0.7 W, 0.15 CF/B | 0.7 W 1.05 CF/B | 61 |
| | 30 | 240 (POAlk) | 4 CH/T, 4 W/MeOH | 4 CH/T, 16 W/MeOH | 83 |
| Fritzsche [17] | ca. 8 | 2.7 (PAN) | 0.1 DMF, 0.08 DHN | — | — |
| V. Tavel et al. [4–8] | 8.85–13.5[c] | 1.5–4 (PMMA) | 0.029 | ca. 6 | 1500–4000 |
| Kuhn [9, 10] | < 10 | 0.4 | 0.4 | — | — |
| Stejskal et al. [13] | 20 | 4 | 0.2 | — | — |

[a] Starting concentration related to the total volume of conjugated phases.
[b] See abbreviations and Table 11.2.
[c] Decreasing $c_0$ with increasing MW.

### 11.1.3 Partition Fractionation of Homopolymers and Chemically Homogeneous Copolymers

A successful partition fractionation requires careful preparatory investigations to find out optimal conditions for partition related to the desired end. We shall discuss here the mode of operation at constant temperature using mostly three or four liquid components.

#### 11.1.3.1 Preparatory Investigations

Preparatory investigations cover (i) the determination of the phase diagram (binodal, tie-lines, critical point) without and with the dissolved polymer, (ii) the derivation of suitable fractionation conditions (composition and temperature of the system), and (iii) the estimation of the distribution coefficients $K_i$ for different polymer species.

**Determination of the Phase Diagram**

Firstly, the miscibility gap in the phase triangle of a ternary solvent system should be determined without the polymer to be fractionated. This can be

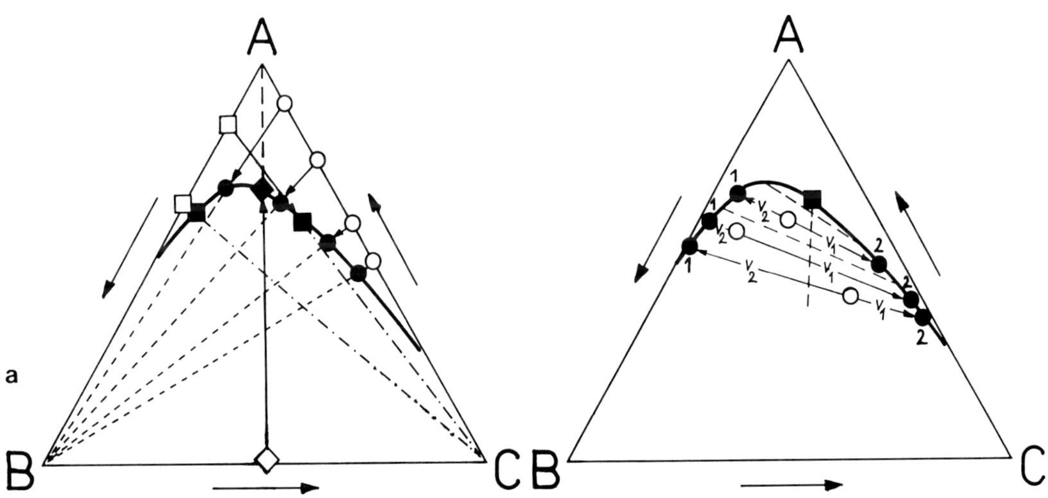


**Fig. 11.4a,b.** Determination of the miscibility gap (binodal) in a ternary system consisting of solvents *A*, *B* and *C* at constant temperature: **a** Turbidimetric-titration or cloud-point measurements starting with mixtures *AB* (□) or *AC* (○) and successive addition of *C* (–·–·–) or *B* (– – –), respectively. Use of an immiscible mixture *BC* (◇) and addition of *A* (– – –) up to the point of first miscibility (◆) is also possible. *Full lines with arrows* symbolize addition of the corresponding solvents. **b** Determination of composition and volume ratio of conjugated phases *1* and *2* (●) originated from immiscible ternary systems (○). *Full lines with arrows* are the tie-lines with directions of demixing, $V_1$ and $V_2$ are the volume fractions of phases *1* and *2* in a corresponding phase pair, respectively. ■ – critical point which is found by connection of the centres of tie-lines and extrapolation to the binodal (– – –)

performed in two ways schematically shown in Fig. 11.4:

- Turbidimetric-titration or cloud-point measurements of binary mixtures by successive addition of the third component yield the binodal covering the miscibility gap (Fig. 11.4a).
- When detailed information about the resulting phases are necessary, demixing experiments with analyses of the phases (composition and volume) must be done (Fig. 11.4b). The results lead to the miscibility gap (binodal) with tie-lines. The centres of the latter render the extrapolation of the critical point.

The phase triangle, applied only to the liquid components of the fractionation system, should be checked with the polymer (different concentration) in the same way.

**Fractionation Conditions**

Composition and volume ratio of the conjugated phases are fixed by composition and temperature of the starting ternary solvent mixture. Each temperature generates its own miscibility gap (cf. Fig. 11.1). In view of the desired $K_i$ values, one must weigh between capacity of polymer solubility, density difference, and asymmetry of partition. Solubility descends with growing distance from critical point whereas the other parameters increase. The volume ratio of the conjugated phases depends upon the position of the starting system on the tie-line. Equal volumes result from starting compositions at the centre of tie-lines (see Fig. 11.4b).

**Estimation of Distribution Coefficients**

Distribution coefficients of samples having graded MW must be measured in separate experiments by evaporation of aliquote volumes of the phases, drying and weighing of the contained polymer. The $K_i$ values obtained can then be used for optimization of the overall composition and temperature of the mixture to yield suitable coefficients. Concentration dependence of $K_i$ should be checked.

### 11.1.3.2 Fractionation Steps

Fractionation steps shall be discussed for two procedures: Simple and counter-current partition. Table 11.4 shows a schematic graph of these variants.

**Simple Partition at Constant Temperature**

**Dissolution of the polymer.**
The polymer to be fractionated should be dissolved in the main component of the mixture, usually a good solvent (A in Figs. 11.1 and 11.4). Elevated temperatures may be necessary for crystalline polymers.
Polymer concentration can be relatively high if necessary (cf. Table 11.3).

**Table 11.4.** Schematic course of partition fractionations at constant temperature

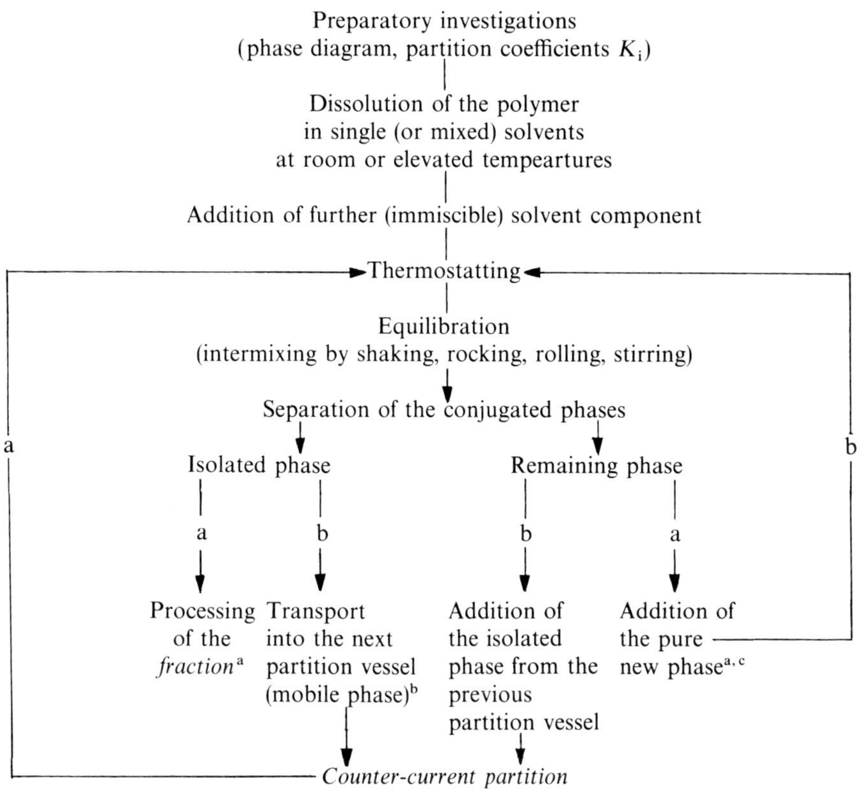


[a] Simple partition procedure.
[b] Counter-current partition – see Table 11.5a–c.
[c] This new phase has mostly the same or a similar composition as the isolated phase.

**Addition of further solvent components.**
Further (immiscible) solvent components are added according to the results of the preparatory investigations (usually under intermixing).

**Thermostatting and equilibration.**
Agitation of the phases for adjustment of the partition equilibrium must proceed at constant temperature. The procedure must be performed carefully. If emulsification occurs, slow rolling of the partition vessel is recommended [1]. Equilibration times vary widely: two hours rolling [1], 30 minutes shaking [3], three minutes shaking [7], 50 times rocking [16]. The time necessary for the actual system can be checked by comparison of the distribution coefficients after different equilibration periods.

**Phase separation.**
After equilibration, phases must separate from each other. The time required can vary from minutes to hours depending on the difference in phase density and on the surface properties (emulsification). Small bubbles on the wall of the separation funnel can be removed by slight vibration [11]. Froth at the interface can be broken by vigorous shaking or swirling of the funnel after the clear (lower or upper) phase has been removed [16]. If phase separation proceeds too slowly or even fails to occur, centrifugation at constant temperature may help [7].

**Phase isolation.**
This step can be implemented similarly to precipitation fractionation and coacervate extraction either with syringes or with the devices shown in Fig. 5.2.

When isolation of the phases is followed by a temperature decrease from equilibration to room temperature, addition of a good solvent to the isolated phase, e.g., use of a syringe partly filled with a good solvent [13], might be necessary in order that the polymer remains soluble.

**Addition of the new phase.**
The phase isolated as a fraction must be replaced, for the following partition step, by an equivalent solvent mixture. This new phase usually has the same volume and composition as the removed phase (without polymer) but slight variations of composition or partition temperature from step to step are also possible.

**Processing of fractions.**
Polymer fractions are obtained as solutions. Therefore, they can be processed either as solution (e.g., in SEC) or, as in the foregoing fractionation techniques, via precipitation (cf. Sect. 4.2.2).

**Counter-Current Partition at Constant Temperature**

**Dissolution of the polymer**
See the same paragraph in simple partition at constant temperature. However, the polymer concentration for counter-current procedures is often lower than for simple partition.

**Addition of further solvent components.**
See simple partition at constant temperature.

**Filling of the separation funnels.**
The volume of each phase should occupy one fourth to one third of the separation funnel. Different variants of filling are possible at the beginning of the counter-current procedure:

(a) Start with the total mass of polymer in only *one* funnel (as reported in Refs. [3, 16]): The polymer is dissolved in the phase mixture. Other funnels are filled with polymer-free mixed phases or only with the stationary phase according to the equilibrium composition. This variant yields the highest fractionation efficiency at equal volume ratio and number of funnels and steps in comparison to other variants (see below). However, only a relatively small amount of polymer can be fractionated.

(b) Start with polymer in *several* funnels (as reported in Ref. [7]): These funnels are filled with the starting mixture containing the dissolved polymer. Other funnels are filled as in variant (a). The fractionation efficiency remains constant in comparison to (a) only by use of additional funnels (as many as filled with polymer). Here, a higher mass of polymer than in (a) can be fractionated under comparable conditions.

(c) Start with *all* funnels filled with the stationary phase according to the equilibrium composition and the polymer dissolved in this phase (similar to counter-current coacervate extraction, cf. Sect. 6.5.2): Mobile phases are added to the stationary phases stepwise during the course of the counter-current procedure starting with the first vessel (see below). A higher polymer mass can be fractionated with an equal number and size of separatory funnels and similar total volume of the solvents than in variants (a) and (b) but the fractionation efficiency worsens.

**Thermostatting and equilibration.**
When the counter-current process is performed automatically (as in a Craig partition experiment), a thermostatted room should be used. Counter-current procedures by hand [3, 16] can be carried out in a normal thermostat. Techniques and time of intermixing are mentioned in variant (i), Ref. [3, 7, 16].

**Phase separation.**
See simple partition at constant temperature.

**Isolation and transport of phases.**
This step can be performed automatically or by use of syringes. Isolation and transport of the mobile phase are immediately connected. The lower phase can also be drained with a stopcock. Notice, however, the potential problems of the tightness mentioned on p. 49.

A quantitative isolation of phases is not necessary. A part of the mobile phase, e.g., one fourth, can remain in the respective vessel. In this case, however, the number of partition steps must be increased in the same ratio, i.e., by the factor 4/3 in the example considered here [7].

The course of the transport process is schematically shown in Tables 11.5a–c. Three variants can be distinguished (see above):

(a) Start with polymer in *one* vessel (Table 11.5a): Vessels $n > 1$ contain pure lower phase ($L_0$). Fractions (underlined) can result from both upper or lower phases.

(b) Start with polymer in *several* vessels (Table 11.5b): The scheme is applied here to three starting vessels.

**Table 11.5a.** Schematic course of counter-current partition fractionation – start with polymer in *one* vessel (restricted to 5 vessels n and IV steps N)[a]

Vessels n 1 2 3 4 5 ...

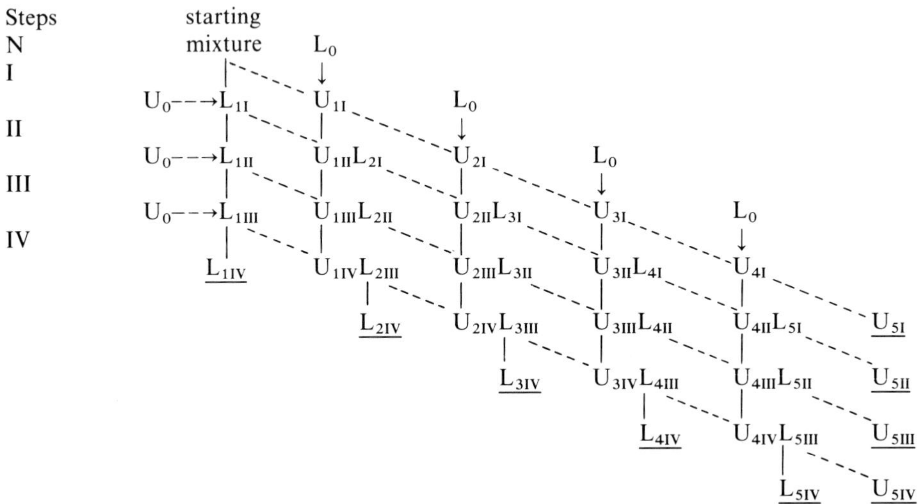


[a] L – lower phase, U – upper phase, subscript $_0$ – pure phase, roman numerals (N) – fractionation step related to the phase, arabic numerals (n) – vessel in which the phase originated, —— – path of lower phases, ---- path of upper phases, fractions are underlined. For details see text.

**Table 11.5b.** Schematic course of counter-current partition fractionation – start with polymer in *three* vessels (restricted to 5 vessels n and IV steps N)[a]

Vessels n 1 2 3 4 5 ...

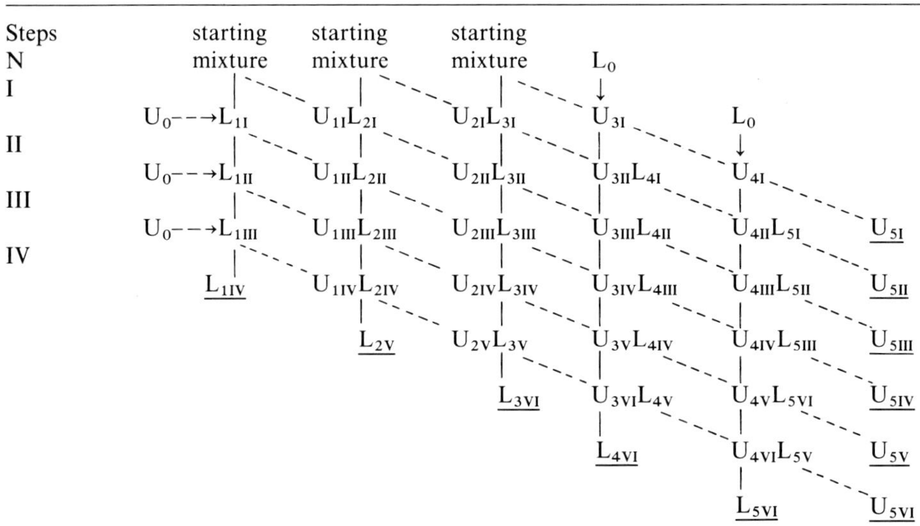


[a] L – lower phase, U – upper phase, subscript $_0$ – pure phase, roman numerals (N) – fractionation step related to the phase, arabic numerals (n) vessel in which the phase originated. —— – path of lower phases, ---- path of upper phases, fractions are underlined. For details see text.

**Table 11.5c.** Schematic course of counter-current partition fractionation – start with polymer in *all* vessels (restricted to 5 vessels n and IV steps N)[a]

| Vessels n | 1 | 2 | 3 | 4 | 5 | ... |
|---|---|---|---|---|---|---|
| Steps N | $U_0 \dashrightarrow L_1$ | $L_2$ | | | | |
| I | $\vert$ | $\downarrow$ | | | | |
| | $U_0 \dashrightarrow L_{1I}$ | $U_{1I}$ | $L_3$ | | | |
| II | $\vert$ | $\vert$ | $\downarrow$ | | | |
| | $U_0 \dashrightarrow L_{1II}$ | $U_{1II}L_{2I}$ | $U_{2I}$ | $L_4$ | | |
| III | $\vert$ | $\vert$ | $\vert$ | $\downarrow$ | | |
| | $U_0 \dashrightarrow L_{1III}$ | $U_{1III}L_{2II}$ | $U_{2II}L_{3I}$ | $U_{3I}$ | $L_5$ | |
| IV | $\vert$ | $\vert$ | $\vert$ | $\vert$ | $\downarrow$ | |
| | $\underline{L_{1IV}}$ | $U_{1IV}L_{2III}$ | $U_{2III}L_{3II}$ | $U_{3II}L_{4I}$ | $U_{4I}$ | |
| | | $\underline{L_{2IV}}$ | $U_{2IV}L_{3III}$ | $U_{3III}L_{4II}$ | $U_{4II}L_{5I}$ | $\underline{U_{5I}}$ |
| | | | $\underline{L_{3IV}}$ | $U_{3IV}L_{4III}$ | $U_{4III}L_{5II}$ | $\underline{U_{5II}}$ |
| | | | | $\underline{L_{4IV}}$ | $U_{4IV}L_{5III}$ | $\underline{U_{5III}}$ |
| | | | | | $\underline{L_{5IV}}$ | $\underline{U_{5IV}}$ |

[a] L – lower phase, U – upper phase, $U_0$ – pure upper phase, $L_n$ – lower phase containing polymer, roman numerals (N) – fractionation step related to the phase, arabic numerals (n) – vessel in which the phase originated, — – path of lower phases, ---- path of upper phases, fractions are underlined. For details see text.

(c) Start with polymer dissolved in the stationary (lower) phase in *all* vessels (Table 11.5c): The procedure works with lower phases $L_n$ containing the dissolved polymer. All $L_n$ have equal composition and volume. The starting mixture in vessel 1 consists of the phases $L_1$ (with polymer) and $U_0$ having the equilibrium composition of solvents.

In a manual counter-current procedure, the numbers of vessels and steps can be chosen arbitrarily whereas both numbers are usually connected in the automated process: n vessels result in n − 1 = N partition steps per vessel. The difference in phase densities is important, especially for the automated transport of the mobile phases [7].

**Processing of fractions.**
Fractions can be isolated as usual (cf. above, Simple Partition) from the consecutive upper phases in the last vessel as well as from the remaining lower phases in each vessel (cf. Tables 11.5a–c).

### 11.1.4 Fractionation of Polymer Blends and Chemically Heterogeneous Copolymers with Demixing Solvents

We will now discuss the fractionation with demixing solvents by temperature decrease of binary solvent mixtures. An extensive compilation of demixing solvent systems reported by Kuhn [11] is given as Table 11.6.

**Table 11.6.** Systems for fractionation of polymer blends with demixing solvents (taken from Ref. [11], with permission)[a, b]

| Polymer fractionated into upper phase | Polymer fractionated into lower phase | Volume fractions of the solvents | Separation temperature (°C) |
|---|---|---|---|
| *cis*-1,4-polybutadiene | Polystyrene | 0.3 DMF, 0.7 CH | 25 |
| *cis*-1,4-polybutadiene | Polystyrene | 0.4 DMF, 0.6 MCH | 25 |
| Polyisobutylene | Polystyrene | 0.24 DMF, 0.76 MCH | 45 |
| Polypentenamer | Polystyrene | 0.2 DMF, 0.8 MCH | 40 |
| Poly-4-methylpentene | Polystyrene | 0.25 DMF, 0.75 MCH | 25 |
| Polyethylene, high density | Polystyrene | 0.25 DMF, 0.75 MCH | 25 |
| Polyethylene, low density | Polystyrene | 0.24 DMF, 0.76 MCH | 25 |
| Polydimethyl siloxane | Polystyrene | 0.4 DMF, 0.6 MCH | 25 |
| Polypropylene | Polystyrene | 0.4 DMF, 0.6 DHN | 45 |
| Polyoxypropylene | Polystyrene | 0.5 DMF, 0.5 MCH | 25 |
| Polyethylhexyl acrylate | Polystyrene | 0.4 DMF, 0.6 MCH | 25 |
| Ethylene (53)/ propylene (47) copolymer | Polystyrene | 0.2 DMF, 0.8 MCH | 25 |
| Ethylene (50)/ propylene (43)/ dicyclopentadiene (7) copolymer | Polystyrene | 0.2 DMF, 0.8 MCH | 25 |
| Ethylene (91.5)/ vinyl acetate copolymer | Polystyrene | 0.3 DMF, 0.7 MCH | 25 |
| Ethylene (60)/vinyl acetate copolymer | Polystyrene | 0.4 DMF, 0.6 MCH | 25 |
| Polydimethyl siloxane | Polymethyl methacrylate | 0.5 DMF, 0.5 MCH | 25 |
| Polyoxypropylene | Polycaprolactone | 0.5 DMF, 0.5 MCH | 25 |
| Polydimethyl siloxane | Polyurethane | 0.4 DMF, 0.6 MCH | 25 |
| Polydimethyl siloxane | Polybutyl acrylate | 0.4 DMF, 0.6 Hp | 25 |
| Polybutadiene | Polybutyl acrylate | 0.4 DMF, 0.6 Hp | 25 |
| Polydimethyl siloxane | Bisphenol-A-polycarbonate | 0.5 DMF, 0.5 MCH | 25 |
| Polydimethyl siloxane | Poly-α-methyl styrene | 0.4 DMF, 0.6 MCH | 25 |
| Polyisobutylene | Poly-α-methyl styrene | 0.32 DMF, 0.68 MCH | 40 |
| *cis*-1,4-polybutadiene | Poly-α-methyl styrene | 0.34 DMF, 0.68 MCH | 40 |
| Polydimethyl siloxane | Poly-4-chlorostyrene | 0.4 DMF, 0.6 MCH | 25 |
| Polydimethyl siloxane | Polytrifluoromethyl styrene | 0.3 DMF, 0.7 MCH | 25 |
| *cis*-1,4-polybutadiene | Polyvinyl chloride | 0.4 DMF, 0.6 MCH | 25 |
| Polydimethyl siloxane | Polyvinyl chloride | 0.4 DMF, 0.6 MCH | 25 |
| Polyethylene, low density | Polyvinyl chloride | 0.25 DMF, 0.75 MCH | 25 |
| Chlorinated polyethylene (chlorine content <36% weight) | Polyvinyl chloride | 0.4 DMF, 0.6 MCH | 25 |

**Table 11.6.** Continued

| Polymer fractionated into upper phase | Polymer fractionated into lower phase | Volume fractions of the solvents | Separation temperature (°C) |
|---|---|---|---|
| Polydimethyl siloxane | Polyoxyethylene | 0.2 DMF, 0.8 MCH | 25 |
| Polydimethyl siloxane | Polyoxyethylene | 0.28 MeOH,0.72 Hp | 25 |
| Polyisobutylene | Polyoxyethylene | 0.2 DMF, 0.8 MCH | 25 |
| Polyethylene, low density | Polyoxyethylene | 0.2 DMF, 0.8 MCH | 25 |
| Polydimethyl siloxane | Polyvinyl acetate | 0.5 DMF, 0.5 MCH | 25 |
| Polydimethyl siloxane | Polyvinyl pyridine | 0.6 DMF, 0.4 MCH | 25 |
| Polydimethyl siloxane | Ethylene (30)/vinyl acetate (70) copolymer | 0.48 MeOH,0.52 Hp | 25 |
| Polyisobutylene | Styrene (72)/butyl acrylate (28) copolymer | 0.2 DMF, 0.8 MCH | 25 |
| Ethylene (53)/ propylene (47) copolymer | Styrene (72)/butyl acrylate (28)/copolymer | 0.2 DMF, 0.8 MCH | 25 |
| Polyoxypropylene | Styrene (75)/acrylonitrile (25) copolymer | 0.5 DMF, 0.5 MCH | 25 |
| *cis*-1,4-polybutadiene | Styrene (75)/acrylonitrile (25) copolymer | 0.43 DMF, 0.57 MCH | 25 |
| Polypropylene | Styrene (75)/acrylonitrile (25) copolymer | 0.4 DMF, 0.6 DHN | 45 |
| Polydimethyl siloxane | Maleic anhydride (50)/isobutylene (50) copolymer | 0.4 DMF, 0.6 MCH | 25 |
| Polydimethyl siloxane | Acrylonitrile (70)/vinyl acetate (20)/styrene (10) terpolymer | 0.55 DMF, 0.45 MCH | 25 |

[a] A similar table compiled by Kuhn is given in Ref. [9].
[b] Numbers in parentheses represent the composition of copolymers.

#### 11.1.4.1 Preparatory Investigations

These investigations include the determination of the phase diagram, of fractionation conditions, and of distribution coefficients. For solvent selection see Sect. 11.1.1.2.

**Determination of the Phase Diagram**

The composition of the phases in a binary demixing solvent system at constant pressure is determined by the temperature. The overall composition of the system influences only the volume ratio of conjugated phases. Starting from homogeneous mixtures at elevated temperatures, the coexistence curve (binodal) can be found in two ways as illustrated in Fig. 11.5:

- Cloud-point measurements of different mixtures AB in a set of sealed ampoules by slow cooling
- Cooling a mixture AB below the cloud-point temperature, separation and analysis of the phases.

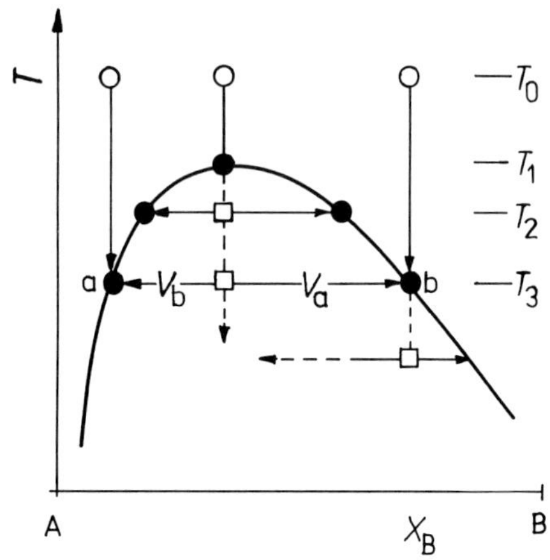


**Fig. 11.5.** Determination of the coexistence curve (miscibility gap) of a binary demixing solvent mixture AB. The coexistence curve can be found either by cloud-point measurements starting with different compositions $X_B$ at elevated temperature $T_0$ (○) or by cooling of a composition $X_B$ to different temperatures $T_i$ (□) and subsequent analysis of the conjugated phases (volume and composition). Phases *a* and *b* (at $T_3$) have volumes $V_a$ and $V_b$, respectively

Analogous investigations are necessary with the system containing the polymer to be fractionated since, due to the polymer, changes of the coexistence curve are possible.

**Fractionation Conditions** The lower the temperature the greater are the differences (composition, density) between conjugated phases and the more complete is the enrichment of one component in one of the phases. The volume ratio of the phases can be controlled by the overall composition of the mixture.

**Estimation of Distribution Coefficients** The dissolved polymer samples (differing in CC and MW) are cooled to different temperatures below the demixing temperature. Then, determination is done generally in the same way as in ternary solvent mixtures (cf. Sect. 11.1.3.1). For copolymers, estimation of MW *and* chemical composition of the polymer in the phases is useful to check the direction of fractionation under the chosen conditions (cf. Sect. 11.1.1.4). In addition, two blank tests each with one of the parent homopolymers should be carried out to see if either homopolymer is concentrated in the different phases.

### 11.1.4.2 Fractionation Steps

Up till now, this fractionation technique has only been performed as a simple stepwise partition. Table 11.7 shows schematically the course of fractionation.

**Dissolution of the Polymer** The heterogeneous copolymer or the mixture of two polymers is dissolved in the mixed solvent at temperatures above the demixing temperature of the mixture. Crystalline polymers must be dissolved above the melting point of crystallites. One should watch out for potential oxidation or degradation by use of elevated fractionation temperatures as discussed in Sect. 4.1.1.

**Table 11.7.** Schematic course of partition fractionation with demixing solvents

Preparatory investigations
(phase diagram, partition coefficients $K_i$)

Dissolution
of polymer blends or heterogeneous copolymers
in mixed solvents
above the demixing temperature

Temperature decrease (demixing)[a]

Thermostatting

Equilibration
(intermixing by shaking, rocking, rolling, stirring)

Separation of the conjugated phases

Isolated phase[b] → Processing of the *fraction*

Remaining phase[b] → Addition of the pure new phase[c] → Heating above the demixing temperature (→ back to Temperature decrease (demixing))

[a] The difference of demixing and equilibration temperature depends on the nature of the copolymer (or polymer mixture) and the desired result (see text).
[b] In the case of one-step fractionations (as possible with polymer mixtures) the conjugated phases are isolated and processed.
[c] This new phase has mostly the same or a similar composition as the isolated phase.

The composition of the mixture must be chosen according to coexistence curve, separation temperature, and desired phase ratio (cf. Fig. 11.5). Polymer concentrations smaller than one percent (e.g., 1 g/500 ml, see [11]) are recommended to suppress effects of incompatibility of polymer mixtures or block and graft copolymers. Nevertheless, polymer concentration up to 20% have been reported [10]. Under certain conditions, a dye-stuff can be added to the homogeneous solution to facilitate the observation of phase separation and isolation. Presuppositions are: The dye-stuff is soluble only in one solvent of the mixture and can be washed out completely from polymer by a polymer nonsolvent. For instance, the DMSO-rich phase in a DMSO/tetrachloroethylene mixture becomes violet by addition of a small amount of fuchsine which can be completely removed by washing the polymer with methanol [13].

**Demixing**

Demixing connected with fractionation is reached by slow cooling the system to the desired separation (equilibration) temperature. Agitation of the system is advantageous. Details of the demixing procedure depend on polymer species:

**Mixtures of two polymers.**
The separation temperature should lie considerably below the demixing temperature because the composition of the phases approaches the pure single solvents with decreasing separation temperature (cf. the coexistence curve in Fig. 11.5). This enhances the efficiency of fractionation since each polymer is almost completely contained in one phase. Therefore, cooling to room temperature is advantageous.

Demixing systems for different polymer pairs are contained in Table 11.6.

**Heterogeneous copolymers.**
Four different fractionation procedures for heterogeneous block, graft, and statistical copolymers were suggested and tested by Kuhn [10, 11]: Summative, successive, semisummative (summative-successive), and "edge" (marginal) fractionation. These variants are shown schematically in Fig. 11.6a–d. The procedures are suitable for measuring CCD. Integral CCDs derived from summative fractionations must be corrected to obtain true distributions [10]. "Edge" fractionation is a tool for approximating the broadness of CCD or for reducing the chemical heterogeneity of the main (middle) fraction.

- Block copolymers [10, 11, 15].

A standard procedure cannot be given because different results has been obtained with different polymers. Kuhn [10] obtained good results for multiple simple demixing of solutions of styrene/butadiene diblock copolymers (summative and successive fractionation) in various mixed solvents with DMF using cooling intervals (for the last fraction) up to more than 40 K. In contrast, Podešva et al. [15] found, with a hydrogenated styrene/isoprene diblock copolymer in DMF/methylcyclohexane, that multiple fractionation steps were only possible with very small successive temperature intervals (separation close to demixing temperature) whereas high temperature differences led to sharp one-step separation. Note that in the fractionation close to the demixing temperature, MWD can perceptibly influence the result. In one-step separation, following by greater intervals, components having chemical compositions above a certain value migrate almost exclusively in one phase while the rest is accumulated in the conjugated phase. Further fractionation of the components requires another solvent system.

- Graft copolymers [13].

The only example described in the literature is the sucessive fractionation of poly(dimethylsiloxane) grafted with methyl methacrylate by demixing in DMSO/tetrachloroethane. The fractionation, starting with the highest temperature difference (about 15 K), is schematically shown in Fig. 11.3. It seems that the extended temperature intervals for individual fractions work successfully because of the rather broad CCD of this copolymer [15].

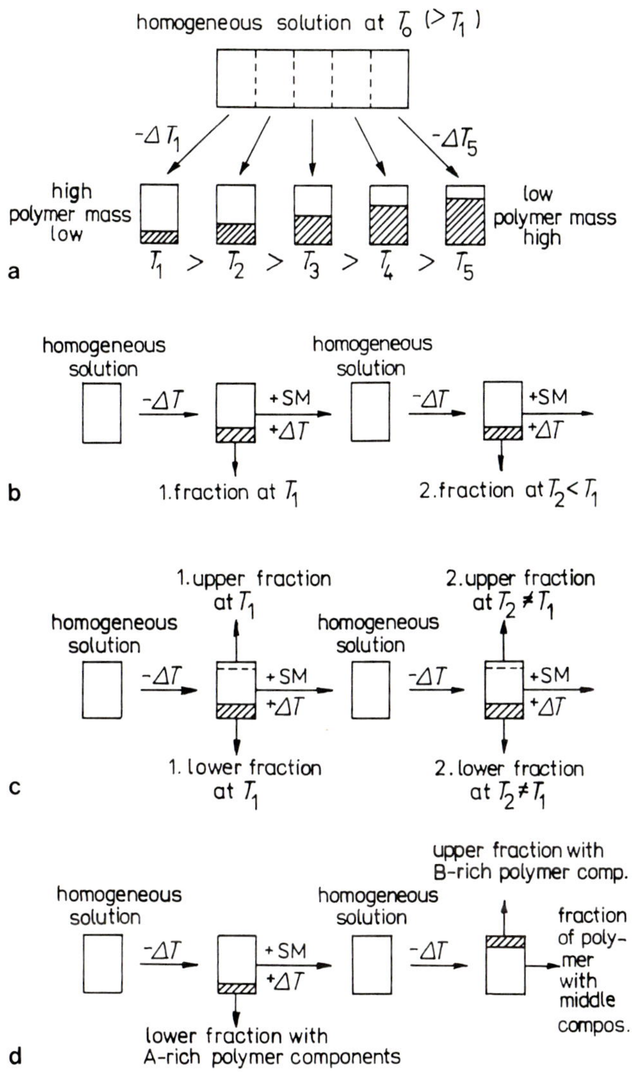


**Fig. 11.6a–d.** Variants of fractionation of heterogeneous copolymers with demixing solvents: **a** – summative fractionation, **b** – successive fractionation, **c** – semisummative (summative–successive) fractionation, **d** – "edge" (marginal) fractionation; *SM* – addition of solvent mixture. (Adopted from Ref. [10] with permission of Hüthig & Wepf Verlag, Basel)

• Statistical copolymers [14].

The fractionation of a styrene/2-methoxyethyl methacrylate copolymer in DMSO/tetrachloroethane is given in detail in Ref. [14]. The same conditions as reported above for a block copolymer [15] were found, i.e., that the multiple fractionation effect could only be achieved in the vicinity of the demixing temperature. In contrast, Kuhn [10] gives the result of a successfully semisummative fractionation of chlorinated poly(ethylene) in DMF/methyl cyclohexane using greater temperature differences.

**Thermostatting and Equilibration** Since the demixing starts with a homogeneous solution, the partition equilibrium adjusts quickly during cooling. Extended thermostatting (usually overnight) is necessary for separation (coagulation) of the phases.

**Phase Separation and Isolation** Generally, the same aspects hold as described for homopolymers (cf. Sect. 11.1.3.2). However for block and graft copolymers, some peculiarities are possible: Block copolymers may show complicated phase behavior due to association effects of the copolymer molecules [15]. Turbid phases can result from fractionation of graft copolymers without impairment of the fractionation efficiency [9–11].

For phase isolation, see p. 160.

**Addition of the New Phase** The new phase which replaces the isolated one must be added (cf. Sect. 11.1.3.2) and the new system must be heated above the demixing temperature to start the next fractionation step by slow cooling. This step is dropped when one-step seperation with isolation of upper and lower phases is carried out.

**Processing of Fractions** See the foregoing fractionation techniques (Sects. 4.2.2, 4.2.3, and 11.1.3.2).

## 11.2 Phase-Distribution Chromatography

In this variant of partition fractionation, the polymer to be fractionated is distributed between a stationary, swollen polymer gel and the polymer solution as the mobile sol phase. The method was systematically developed by Casper and Schulz [22] and by Greschner [23]. Some first elements of the method are reported in Ref. [19].

**Principle of the Method** The high-MW polymer phase is generated by precipitation on glass beads (similar to column fractionation procedures described in the foregoing sections) and swelling of the polymer in a theta solvent below the theta temperature of the system (cf. Glossary). Low-MW portions of this stationary gel phase are extracted with the theta solvent at a temperature slightly below the theta temperature. The adherence of the swollen gel phase to the support is essential. Therefore, fractionations are carried out at least 10 K below the theta temperature. The theta solvent in which the polymer solution is injected on top of the column works as a mobile sol phase at constant temperature and with constant flow rate. The method acts without a gradient.

The MW of the polymer to be fractionated must be smaller than MW of the stationary phase by a factor 10 at least. Distribution coefficients $K_i = c''/c'$ (cf. Eq. (3.8)) increase with decreasing temperature and with increasing MW. Hence, low-MW fractions leave the column first. Calibration of the method (elution volume vs. MW) is possible with narrow standard samples. $K_i$ is, in a wide range, almost independent of polymer concentration.

Hitherto, only polymers with the same chemical structure have been used as the gel phase and as sample. Otherwise, problems may result from incompatibility of different polymers. Phase-distribution chromatography has been used in analytical scale to determine MWDs precisely.

**Application**

Use of this method has only been reported for poly(styrene): Casper and Schulz [22] used PS (MW about $10^7$ g/mol) precipitated on glass beads (ca. 0.1 mm) as the gel phase in a 100 cm × 3 cm column. This phase was extracted with cyclohexane at 28 °C (theta temperature 35 °C). The insoluble film swelled to a thickness of 300 to 500 nm yielding a volume ratio $V'/V''$ of about 30. After this preparation of the column, up to 350 mg samples of PS (MW < $10^6$ g/mol, dissolved in 100 ml cyclohexane) were injected into a cyclohexane stream of about 10 ml/h. Elution temperatures were below 25 °C.

An automated apparatus working on the principle described by Casper and Schulz was developed by Greschner [23]. He used glass beads of 70 to 80 µm diameter and a V4A steel column with a total length of 584 cm and an inner diameter of 1 cm. The thickness of the PS film was 305 nm and cyclohexane flowed at 15 ml/h. Calibration was made with PS standards. The apparatus makes possible the ascertainment of very small differences in MWDs of PS.

## 11.3 Examples of Partition Fractionations

### 11.3.1 Partition Fractionation of Poly(oxyethylene) with Water/Chloroform/Benzene

This example is designed for 1 g polymer. The given specifications for volumes of vessels and solvents can be converted linearly for other polymer masses. The fractionation can be performed as simple or counter-current partition. Owing to the use of small amounts of benzene, the procedure must be carried out completely in an exhauster.

#### 11.3.1.1 Simple Partition Fractionation

**Materials**

**Polymer.** Poly(oxyethylene) (POE), 1 g, MW about $10^4$ g/mol

**Solvents.** Water (W), 100 ml;
Chloroform (CF) (analytical grade), 35 ml (related to five fractions);
Benzene (B) (analytical grade), 20 ml (related to five fractions)

**Nonsolvent.** Cyclohexane or *n*-hexane, ca. 1 l for precipitation of the polymer (1 g totally)

Materials required for preparatory tests are not included.

**Equipment**

- Standard glassware: conical flasks (100 ml), glass beakers (100 ml), flasks with ground-glass joints (100 ml)
- Syringes or pipettes (0–50 ml)
- Shaking machine or similar device
- Thermostat (basin)
- Rotational-type evaporator
- Separation funnels (slender form) with ground-glass joint and stopper (100 ml)

Equipment for re-dissolution and re-precipitation of the fractions is not considered here.

**Time Required**

| | |
|---|---|
| – Dissolution of the polymer | 1–2 h |
| – Equilibration | 0.5 h |
| – Phase separation | $\leqslant$2 h |
| – Phase isolation and addition of the new phase | 0.5 h |

All times except the first one are valid in each extracting step. Times for processing of phases and polymer fractions are not included.

**Preparatory Investigations.** The limits of partition fractionation ($K_i = 1$ or $\infty$) should be estimated in the following way at 25 °C:

- Partition of equal masses POE in phase mixtures having different CF/B ratio ($c_P$ ca. 10 g/l; W/(CF + B) = 10:1 (v/v); CF/B (v/v) = 10:0, 9:1, . . ., 1:9, 0:10)
- Determination of masses in the conjugated phases. Fractionation is expected for contents of B between the values leading to $K_i = 1$ and $\infty$, respectively. Perhaps, in a second series, the CF/B ratio must be graded for certain mixtures in smaller steps.

**Fractionation** Fractionation temperature: 25 °C

- Dissolution of 1 g POE in 100 ml W
- Addition of 10 ml CF/B mixture with about 50% B (cf. with preparatory investigations)
- Thermostatting and equilibration by gentle rocking (about 50 times)
- Isolation of the organic phase after phase separation
- Addition of 10 ml organic phase containing a smaller amount of B; the content of B diminishes with increasing number of the fractionation steps (differences depend on number of fractions, lowest value about 20% B); after thermostatting and equilibration, each organic phase is isolated and replaced by the next CF/B mixture
- Evaporation of the solvents in the organic phase to isolate the polymer. The concentrated solution can also be precipitated into cyclohexane or *n*-hexane in excess.
- Processing of the fraction as usual.

**Evaluation**

- Determination of mass and MW or an equivalent value ($[\eta]$, $V_e$ from SEC, $\varphi^*$) of the fractions
- Deduction of conclusions for optimization of the procedure (mass balance and MW gradation of the fractions; repeat with variation of CF/B ratio in the individual steps or different ratios W/organic phase)
- Comparison with the result of Sect. 11.3.1.2.

### 11.3.1.2 Counter-Current Partition Fractionation

**Materials**

See Sect. 11.3.1.1. The required volumes of the solvents differ from the previous example. For a partition with four steps and five vessels, the following estimated volumes are necessary: 500 ml W, 30 ml CF, 20 ml B.

**Equipment**

Because a simple variant of a counter-current partition is to be carried out, the equipment from Sect. 11.3.1.1 can be used.

**Time Required**

The temporal course of one partition step is similar to the simple partition described in Sect. 11.3.1.1.

**Preparatory Investigations**

According to the results of Sect. 11.3.1.1, an intermediate B content should be chosen (perhaps about 40%).

**Fractionation**

Fractionation temperature: 25 °C

- Dissolution of 1 g POE in 100 ml W
- Addition of 10 ml CF/B mixture with about 40% B
- Thermostatting and equilibration by gentle rocking (about 50 times)
- Transportation of one phase in the next vessel and replacement of the transportet phase by the same one; addition of the conjugated phase in the second vessel, etc. The phase composition keeps constant with respect to liquid components. The scheme is given in Table 11.5a. Different modifications are possible:

● The upper (water-rich) phase is transported (as in Table 11.5a).
● The lower (CF-rich) phase is transported (symbols L and U in Table 11.5a are reversed).
● Both variants can be performed with varying steps N per vessel.

- Isolation of lower and upper phases as fractions according to the scheme in Table 11.5a
- Isolation of the polymer fractions by evaporation or precipitation and processing analogously to Sect. 11.3.1.1

**Evaluation**

See Sect. 11.3.1.1. Comparison is possible with Sect. 11.3.1.1 and for the counter-current variants recommended above. For optimization, repeat using other CF/B or W/organic-phase ratios is possible.

### 11.3.2 Partition Fractionation of Poly(styrene)/Poly(butadiene) Mixtures by Demixing in DMF/Methylcyclohexane[1]

All specifications of this example are referred to 500 mg polymer to be fractionated. Instead of poly(butadiene), poly(dimethyl siloxane) can also be used.

**Materials**

**Polymers.** Poly(styrene) (PS) 750 mg;
Poly(butadiene) (PBD), 750 mg

**Solvents.** DMF, ca. 300 ml;
Methylcyclohexane (MCH), ca. 400 ml

**Nonsolvents.** Methanol/acetone mixture (about 1:1) for precipitation tests, about 100 ml; for complete precipitation of the polymers, about 1 l nonsolvent mixture is necessary.

**Equipment**

- Three-neck flask equipped with tight-fitting stirrer and reflux condenser (alternatively, conical flask with magnetic stirrer and reflux condenser may be used)
- Heater (bath or hot plate)
- Separation (dropping) funnel (250–500 ml) furnished with heating jacket
- Syringes or pipettes
- Rotating-type evaporator with 250-ml flasks
- Glass beakers (100 ml, 500 ml)
- Stirrer
- Small dropping funnel
- Glass rod
- Device for filtration by pressure or vacuum
- Vacuum oven
- Stands, clamps, sockets, standard laboratory glassware

**Time Required**

- Dissolution of the polymer 1–2 h
- Demixing by cooling ca. 1 h
- Phase separation ca. 1 h
- Phase isolation 0.5 h

Times for processing of phases and polymer fractions are not included.

**Preparatory Investigations**

- Study of the demixing behaviour of the solvent mixture in dependence of composition and temperature

[1] This example was prepared using information kindly given by Dr. R. Kuhn. This help is gratefully acknowledged.

- Miscibility test for solvents and nonsolvents (Why is methanol/acetone used instead of pure methanol?)
- Demixing experiments using one polymer only (carrying-out analogously to Fractionation using 500 mg of only one polymer)

**Fractionation**

- Dissolution of 250 mg PS and 250 mg PBD into 80 ml DMF and 120 ml MCH by refluxing with stirring at about 80 °C
- Transportation of the hot solution into the separation funnel which is thermostatted at 75 °C (the solution must remain clear)
- Slow cooling to 25 °C
- After phase separation: taking of a few milliliters of each phase in about 50 ml nonsolvent mixture, respectively (comparison with preparatory investigations?)
- Isolation of the phases
- Evaporation of solvents or precipitation (after partial evaporation) into the nonsolvent mixture
- Filtration and drying

**Evaluation**

- IR analyses of the polymer fractions and comparison with the results of preparatory investigations
- Comparison of applied and isolated polymer masses

## References


1. Schulz GV, Nordt E (1940) J Prakt Chem, NF 155: 115
2. Schulz GV (1940) Z Phys Chem B 46: 137
3. Almin KE (1957) Acta Chem Scand 11: 1541; (1959) Acta Chem Scand 13: 1263, 1274, 1278, 1287, 1293
4. v Tavel P, Bieri V (1971) Makromol Chem 149: 63
5. v Tavel P (1972) Chimia 26: 187
6. v Tavel P, Wälchli J, Rüfenacht H (1973) Chimia 27: 86
7. v Tavel P, Rüfenacht HJ (1976) Makromol Chem 177: 2431
8. Rüfenacht HJ, v Tavel P (1976) Makromol Chem 177: 2449
9. Kuhn R (1976) Makromol Chem 177: 1525
10. Kuhn R (1980) Makromol Chem 181: 725
11. Kuhn R (1983) In: Klempner D, Frisch KC (eds) Polymer alloys III. Plenum, New York, p 45
12. Kuhn R, Müller HG, Bayer G, Krämer-Lucas H, Kaiser W, Orth P, Eichenauer H, Ott KH (1993) Colloid Polym Sci 271: 133
13. Stejskal J, Straková D, Kratochvil P, Smith SD, McGrath E (1989) Macromolecules 22: 861
14. Podešva J, Procházka O, Stejskal J, Špaček P, Enders S (1993) J Appl Polym Sci 48: 1127
15. Podešva J, Stejskal J, Kratochvil P (1993) J Appl Polym Sci 49: 1265
16. Case LC (1960) Makromol Chem 41: 61
17. Fritzsche P (1971) Faserforsch Textiltechn 22: 535
18. Hamann K, Funke W, Gilch H (1959) Angew Chem 71: 596

19. Glöckner G (1987) Polymer characterization by liquid chromatography. Elsevier, Amsterdam
20. Englert A, Tompa H (1970) Polymer 11: 507
21. Albertsson PA (1970) Adv Protein Chem 24: 309
22. Casper RH, Schulz GV (1970) J Polym Sci A-2, Polym Phys 8: 833; (1971) In: Altgeld KH, Segal L (eds) Gel permeation chromatography. Marcel Dekker, New York, p 225
23. Greschner GS (1979) Makromol Chem 180: 2551

# 12 Cross Fractionation

## 12.1 Aims and Principles

Copolymers generally possess a two-dimensional distribution: MWD and CCD. An analogous problem exists with polyolefines: MWD and distribution of short-chain branching (SCB, cf. Sect. 10).

Although, as described in Sect. 4.4, various solvent/nonsolvent combinations often favour different directions of fractionation, ascertainment of the distribution of one parameter independently of the second one is very rare. Fractionation according to CC without any influence of MWD is not possible (cf. Eq. (4.7)).

That means for heterogeneous copolymers:

- Determination of the "true" MWD or CCD by fractionation in only one solvent/nonsolvent system is normally impossible. Even when two solvent/nonsolvent combinations yield identical distributions, this is not evidence of the "true" distribution (since these systems may have the same characteristics).
- Fractionation in only one direction (e.g., according to MW) produces fractions having great inhomogeneities of the second parameter (e.g., CC). Thus, fractions with both narrow MWD and CCD cannot be obtained.

The problems can be overcome by cross fractionation. Determination of the complex distribution can be performed using analytical fractionation procedures whereas fractions for further use and narrow in both MWD and CCD require preparative fractionations.

Cross fractionation is the consecutive fractionation of a heterogeneous polymer with systems which act in different directions. This necessity was first recognized by Rosenthal and White in 1952 in the context of fractionations of cellulose acetate [1]. Figure 12.1, from Rosenthal and White [1], illustrates the problem:

The polymer with its complex distribution (a) can be fractionated according to MW (b) as well as to CC (c). These variants are very unusual. Usually, solvent/nonsolvent combinations fractionate according to both MW and CC, but in different directions (d, e). All fractions obtained with systems (b) through (e) are rather heterogeneous, at least in one direction. This problem cannot be overcome by a fine-graded fractionation (f). Fractions narrower in

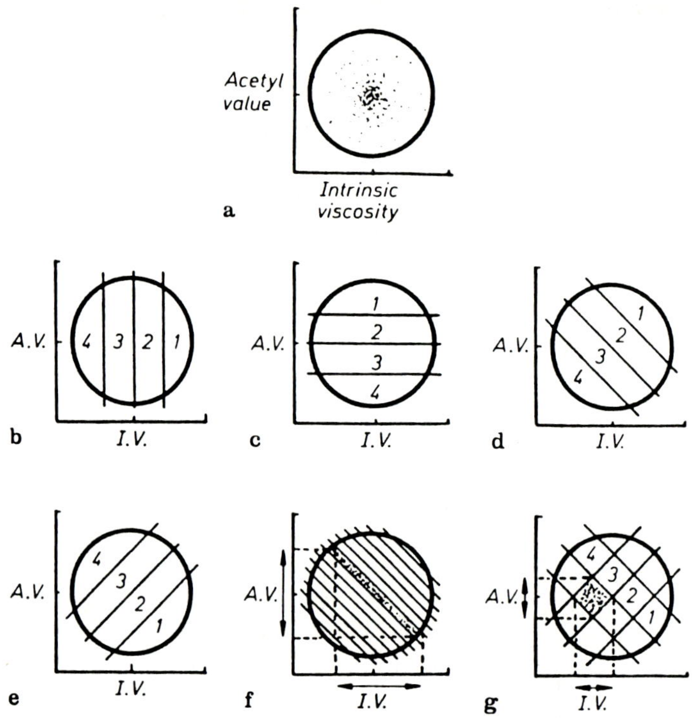


**Fig. 12.1a–g.** Schemes of fractionation and cross fractionation: **a** Complex distribution of acetyl content (*ordinate*) and intrinsic viscosity as a measure of MW (*abscissa*) of a cellulose acetate sample; **b** fractionation according to MW; **c** fractionation according to CC; **d** and **e** general course of fractionations simultaneously according to MW and CC, directions depend upon the solvent/nonsolvent system used; **f** fine-graded fractionation using the combination of Fig. 12.1d; **g** cross fractionation with consecutive use of the system of Fig. 12.1d and the complementary system of Fig. 12.1e. (Reprinted with permission from Ref. [1]. Ind. Engin. Chem. 44 (1952). Copyright 1952 American Chemical Society.)

both MWD and CCD are only available by cross fractionation with two complementary solvent/nonsolvent systems (g). Note, it is not necessary to use systems as in (b) and (c) fractionating parallel to the axes. The best condition is a right angle between the directions of fractionation. In Fig. 12.1g, each fraction obtained in the first fractionation with a system used in (d) was fractionated with a system represented in (e).

Cross fractionation means in terms of Eqs. (4.7) and (4.8) that complementary solvent/nonsolvent combinations change the sign of $\kappa$ since $\chi_{iA} > \chi_{iB}$ and $\chi_{iA} < \chi_{iB}$ must hold for these combinations, respectively.

Generally, each fractionation procedure can be used when the two consecutive methods act in different directions as shown in Fig. 12.1. The sequence of the chosen procedures is unimportant. Results of cross fractionations may be plotted three-dimensionally as "mountains" or two-dimensionally as contour-line maps.

## 12.2 Applications

Let us distinguish three variants of cross fractionation:

- "Classical" cross-fractionation using two different solvent/nonsolvent combinations,
- "non-classical" cross-fractionation combining mostly a fractionation by solubility differences with SEC, and
- chromatographic cross-fractionation coupling two chromatographic methods. This variant, which requires only negligible amounts of polymer (ca. 1 mg) and small amounts of solvent and time, will not be considered here. Comprehensive descriptions are given by Glöckner [2, 3].

### 12.2.1 "Classical" Cross-Fractionation

Table 12.1 quotes a compilation of "classical" cross-fractionations [3]. Precipitation fractionation was used in all cases and also in an example given additionally in the literature [34]: cross fractionation of chlorinated natural rubber with chloroform/*iso*-amyl alcohol (system 1, 4–6 fractions) and benzene/acetone : methanol (1 : 1) (system 2, 12–15 final fractions). The search for solvent/nonsolvent systems suitable for cross fractionations can be performed via turbidimetric titrations or test fractionations of copolymers in different combinations as described in detail in Sect. 4.4.

Owing to a high expenditure of time and materials, "classical" cross-fractionation is only seldom performed although this procedure is irreplaceable, as yet, for the production of fractions with narrow distributions in both MW and CC on a preparative scale.

### 12.2.2 "Non-classical" Cross-Fractionation

The consumption of time and materials can be reduced when one of the fractionation procedures is substituted by a chromatographic method. Hitherto, SEC has been used to separate according to hydrodynamic volume, $[\eta] \cdot M$, which is an equivalent measure of MW when copolymers are investigated having a CCD that is not too broad. SEC can be combined with different solubility methods. Table 12.2 summarizes examples given in the literature.

Cross fractionations with SEC separation yield analytical results. Sometimes, fractions of the first fractionation procedure (if it is not SEC) may be obtained on a preparative scale. Combinations of SEC and turbidimetric titration [17, 18] require only very small amounts of polymer (about 10 mg).

A relatively widely used combination is TREF (see Sect. 10) and SEC yielding the complex distribution of SCB and MW. Results of this technique, which can be performed off-line or on-line, are summarized in Refs. [3,

**Table 12.1.** "Classical" cross-fractionation of two solvent/nonsolvent systems (taken from Ref. [3] with permission)

| Sample | System 1 | | | System 2 | | | Reference |
|---|---|---|---|---|---|---|---|
| | Solvent/ nonsolvent | Direction of separation[a] | No. of intermediate fractions | Solvent/ nonsolvent | Direction of separation[a] | No. of final fractions | Year/Ref. |
| CA | Ac/W | $[\eta]\searrow$, CA $\searrow$ | 5 | Ac/Pn | $[\eta]\searrow$, CA $\nearrow$ | 4[b] | 1952 [1] |
| VAC/N-VP[c] | Ac/EtE | $[\eta]\searrow$, VAC $\nearrow$ | 3 | *i*-PrOH/Hx | $[\eta]\searrow$, VAC $\searrow$ | 12 | 1967 [4] |
| S/BD | B/MEK | S $\nearrow$, $M\searrow$ | 4 | CH/iOct | S $\searrow$, $M\searrow$ | 20 | 1970 [5] |
| S/MMA(az)[d] (15 g) | Ac/AcN | (MMA $\nearrow$) | 6 | BuCl/CH | $M\searrow$ | 35 | 1971 [6] |
| S/AN(15 g) | MEK/CH | AN $\searrow$ | 5 | EtCb/EtCN[e] | AN $\nearrow$ | 32 | 1974 [7] |
| S/MEMA[f] | B/CH | S $\nearrow$ | 5 | MEK/MeOH | S $\searrow$ | 22 | 1978 [8] |
| S/MEMA (az)[d] | B/CH | S $\nearrow$ | 6 | MEK/MeOH | (S $\searrow$) | 27 | 1981 [9] |
| S/MA (5.06 g) | CF/CH | S $\nearrow$ | 4 | HEMeE[g]/MeOH | S $\searrow$ | 19 | 1981 [10] |
| PET/PTME[h] (15 g) | TeCM/MeOH | PET $\searrow$ | 6 | TCE/Hp | PET $\searrow$ | 18 | 1989 [11] |
| MMA/MVK[i] | DMF/W + 0.5% $NH_4Cl$ | $[\eta]\searrow$ | 3 | CF/EtE | MMA $\nearrow$, $[\eta]\searrow$ | 15 | 1977 [12] |
| VAL/VEtCb[j] | THF/Hx | VAL $\searrow$ | 4 | MeOH/W + 1% NaCl | VAL $\nearrow$, $M\searrow$ | 18 | 1983 [13] |
| VAL/VAC[k] | THF/Hx | $M\searrow$ | 5 | MeOH/W + 0.5% NaCl | $M\searrow$, VAC $\nearrow$ | 18 | 1986 [14] |
| VAC/VAL[l] | DMF/T + Hx (1:2) | $M\searrow$ | 4 | Ac + W(3:2)/ W + 0.5% NaCl | $M\searrow$, VAC $\searrow$ | 16 | 1986 [14] |

*Local abbreviations*:
[a] on reduction of dissolution power.
[b] fractionation of the second fraction obtained by system 1.
[c] *N*-VP: *N*-vinyl pyrrolidone.
[d] (az): azeotropic copolymer.
[e] EtCb: ethylene carbonate: EtCN: ethylene cyanohydrine.
[f] model mixture, 10 characterized copolymers.
[g] HEMeE: 2-hydroxyethyl methyl ether.
[h] poly(ethylene terephthalate)/poly(tetramethylene ether) multiblock copolymer.
[i] MVK: units of PMMA, $\beta$-functionalized by $(CH_3)_2NSO_2CH_2Li$.
[j] VEtCb: vinyl ethyl carbonate, by modification of PVAL with ethyl chloroformate.
[k] by partial acetylation of PVAL.
[l] by partial alkaline hydrolysis of PVAC.

**Table 12.2.** "Non-classical" cross-fractionation

| Polymer | Method 1 | | | Method 2 | | | Reference |
|---|---|---|---|---|---|---|---|
| | Procedure | Solvent/ nonsolvent | Direction of separation | Procedure | Solvent/ nonsolvent | Direction of separation | Year/Ref. |
| AN/EtA | partition | DMF/DHN | EtA ↗ | SEC | DMF | $M$[a] | 1971 [15] |
| S/AN | gradient elution | DMF/T + PrOH (1:1) | AN ↗, ([$\eta$] ↗) | SEC | DMF + 0.02 M LiBr · $H_2O$ | $M$[a] | 1981 [16] |
| S/BD/CP[b] | SEC | THF | $M$[a] | TT | THF/MeOH | CC | 1977 [17] |
| S/AN | SEC | THF | $M$[a] | TT | THF/Hx | CC | 1985 [18] |
| LDPE | gradient elution | xylene/ethyl cellosolve | $M$ ↗ | TREF | xylene | SCB | 1965 [19] |
| LDPE | TREF | oDCB | SCB | SEC | oDCB (140 °C) | $M$[a] | 1981 [20] |
| LDPE, LLDPE | TREF | xylene | SCB | SEC | TCB (145 °C) | $M$[a] | 1982 [21] |
| LLDPE | TREF | oDCB | SCB | SEC | oDCB (140 °C) | $M$[a] | 1986 [22] |
| LLDPE | TREF | αCN | SCB | SEC | | $M$[a] | 1987 [23] |
| LDPE, LLDPE | TREF | TCB | SCB | SEC | TCB (140 °C) | $M$[a] | 1987 [24] |
| E/P[c], LDPE | TREF | oDCB | SCB | SEC | oDCB (140 °C) | $M$[a] | 1988 [25] |
| LDPE, LLDPE | TREF | xylene | SCB | SEC | TCB (135 °C) | $M$[a] | 1989 [26] |
| LLDPE | TREF | xylene | SCB | SEC | oDCB (135 °C) | $M$[a] | 1993 [27] |
| Co-graft cp[d] | demixing solvents | DMF/MCH | CE + PVAC/ EVA + EVA – VAC + EVA – VAC-CE | demixing solvents | butyrolactone/ CH/perchloro ethylene | EVA + EVA – VAC/EVA – VAC – CE | 1985 [28] |

[a] The directly operative measure is the hydrodynamic volume which is nearly proportional to MW.
[b] Cyclopentene grafted on S/BD copolymer.
[c] Ethylene/propylene block copolymer.
[d] Co-graft copolymer EVA – VAC – CE (cellulose ester), separated from byproducts in two steps.

29–31]. A remarkable combination to obtain the complex SCB distribution was used by Shirayama et al. as early as 1965 [19] (gradient elution followed by TREF). Recently, a cross fractionation of LLDPE has been reported which was performed as "dynamic direct-extraction" (see p. 72) using mixtures of *p*-xylene and ethylene glycol monoethyl ether (MW fractionation) followed by fractionating solution crystallization (according to SCB, cf. TREF) in *p*-xylene at different temperatures [32].

A special kind of cross fractionation was created by Kuhn and co-workers [28, 33]. They used several mixtures of demixing solvents having different partition characteristics for various polymers to separate a polymer mixture (of more than two constituents) after grafting reactions.

## References

1. Rosenthal AJ, White BB (1952) Ind Engin Chem 44: 2693
2. Glöckner G (1989) In: Booth C, Price C (eds) Comprehensive polymer science, vol 1. Pergamon, Oxford, p 313
3. Glöckner G (1991) Gradient HPLC of copolymers and chromatographic cross-fractionation. Springer, Berlin Heidelberg New York
4. Agasandyan VA, Kudryavtseva LG, Litmanovich AD, Shtern VYa (1967) Vysokomol Soedin A 9: 2634
5. Teramachi S, Kato Y (1970) J Macromol Sci, Chem A 4: 1785
6. Teramachi S, Kato Y (1971) Macromolecules 4: 54
7. Teramachi S, Fukao T (1974) Polymer J 6: 532
8. Stejskal J, Kratochvíl P (1978) Macromolecules 11: 1097
9. Stejskal J, Kratochvíl P, Straková D (1981) Macromolecules 14: 150
10. Teramachi S, Hasegawa A, Hasegawa S, Ishibe T (1981) Polymer J 13: 319
11. Xu Z, Yuang P, Zhong J, Jiang E, Wu M, Fetters LJ (1989) J Appl Polym Sci 37: 3195
12. Bourguignon JJ, Bellissent H, Galin JC (1977) Polymer 18: 937
13. Arranz F, Sanches-Chaves M, Molinero A, Martinez R (1983) Makromol Chem, Rapid Commun 4: 297
14. Arranz F, Sanches-Chaves M, Riofrio A (1986) Makromol Chem 187: 1215
15. Fritzsche P, Klug P, Gröbe V (1971) Faserforsch Textiltechn 22: 250
16. Ogawa T, Sakai M (1981) J Polym Sci A-2, Polym Phys 19: 1377
17. Hoffmann M, Urban H (1977) Makromol Chem 178: 2683
18. Glöckner G, Albrecht V, Francuskiewicz F, Ilchmann D (1985) Angew Makromol Chem 130: 41
19. Shiramaya K, Okada T, Kita SI (1965) J Polym Sci A 3: 907
20. Nakano S, Goto Y (1981) J Appl Polym Sci 26: 4217
21. Wild L, Ryle TR, Knobeloch DC, Peat IR (1982) J Polym Sci A-2, Polym Phys 20: 441; Wild L, Ryle TR, Knobeloch DC (1982) Polym Prepr, Am Chem Soc, Polym Chem Div 23: 133
22. Usami T, Gotoh Y, Takayama S (1986) Macromolecules 19: 2722
23. Keluski EC, Elston CT, Murray RE (1987) Polym Engin Sci 27: 1562
24. Mirabella FM, Ford EA (1987) J Polym Sci A-2, Polym Phys 25: 777
25. Gotoh Y, Usami T, Takayama S (1988) Poster presentation, 1st ISPAC Meeting, Toronto, June 2–3

26. Brauer E, Gebauer E, Wiegleb H, Führling W (1989) Plaste & Kautschuk 36: 9
27. Zhou XQ, Hay JN (1993) Eur Polym J 29: 291
28. Alberts H, Bartl H, Kuhn R, Morbitzer L (1985) Kautschuk + Gummi, Kunststoffe 38: 689
29. Wild L (1990) Adv Polym Sci 98: 1
30. Glöckner G (1990) J Appl Polym Sci, Appl Polym Symp 45: 1
31. Vela Estrade JM, Hamielec AE (1993) Polym Reaction Engin 1: 171
32. Springer H, Hengse A, Hinrichsen G (1993) Coll Polym Sci 271: 523
33. Kuhn R (1983) In: Klempner D, Frisch KC (eds) Polymer alloys III. Plenum, New York, p 45
34. Krentsel LB, Travin SO, Litmanovich AD, Yutujan KK (1985) Eur Polym J 21: 405

# 13 Outlook

Let us finally consider some possibilities of further developments of fractionation techniques.

Analytical fractionations based on differences in solubility of the individual polymer constituents have lost their importance for MWD determination in favour of more efficient methods as SEC [1–3], different field-flow fractionation techniques [4], and modern applications of analytical ultracentrifugation [5]. However, ascertainment of CCD will also be, in addition to modern HPLC techniques [6], a matter of fractionations with demixing solvents in the future (cf. Sect. 11).

On a preparative scale, various fractionation techniques considered in this book remain important. This applies to precipitation fractionation in relation to preparative cross-fractionations (cf. Sect. 12) as well as to uncomplicated production of preparative fractions.

In the field of extraction fractionation, CPF (see Sect. 6.6.2) will be the method of the future. In addition to being perfected, optimized, and automated further [7, 8], an increase in the efficiency of the procedure is imaginable by coupling several columns in series to obtain more than two fractions (sol and gel) per run (e.g., gel(i) flows directly in the next column as feed (i + 1)).

Preparative gradient-elution fractionation (cf. Sects. 7 and 9.2) and TREF (Sect. 10) could be optimized and more automated for routine fractionations.

Advances in partition fractionation can be expected above all in fractionation with demixing solvents. Combination of this principle of fractionation and CPF technique for separation of polymer mixtures (e.g., as a recycling step) is under discussion [9].

In the field of cross fractionation, one will strive for a combination of methods more efficient than precipitation fractionation on a preparative scale. Perhaps, use of CPF technique will also be possible in future for this end.

Thus, preparative fractionation methods remain as important now as ever.

## References

1. Tung LH, Moore JC (1977) In: Tung LH (ed) Fractionation of synthetic polymers. Marcel Dekker, New York, p 545

2. Glöckner G (1987) Polymer characterization by liquid chromatography. Elsevier, Amsterdam
3. Dawkins JV (1989) In: Booth C, Price C (eds) Comprehensive polymer science, vol 1. Pergamon, Oxford, p 231
4. Gunderson JJ, Giddings JC (1989) In: Booth C, Price C (eds) Comprehensive polymer science, vol 1. Pergamon, Oxford, p 279
5. Lechner MD, Mächtle W (1991) Makromol Chem 192: 1183; (1992) Makromol Chem, Macromol Symp 61: 165; (1992) Makromol Chem, Rapid Commun 13: 555; Ortlepp B, Panke D (1992) Makromol Chem, Makromol Symp 61: 176; Ebert G (1992) Makromol Chem, Makromol Symp 61: 185
6. Glöckner G (1991) Gradient HPLC of copolymers and chromatographic cross-fractionation. Springer, Berlin Heidelberg New York
7. Wolf BA (1992) Makromol Chem, Macromol Symp 61: 244
8. Weinmann K, Wolf BA, Rätzsch MT, Tschersich L (1992) J Appl Polym Sci 45: 1265
9. Wolf BA (1991) Hamburger Makromolekulares Symposium, 25.–27.09.1991, Hamburg, FRG

# Appendix

**Table A 1.** Calculation of solubility parameters according to the increment method proposed by Small[a], $\delta = (D/M_0) \cdot \Sigma F$[b]

| Group | F $[(J/cm^3)^{1/2}\ cm^3\ mol^{-1}]$ | Group | F $[(J/cm^3)^{1/2}\ cm^3\ mol^{-1}]$ |
|---|---|---|---|
| $-CH_3$ | 437 | –O– (ethers) | 143 |
| $-CH_2-$ | 272 | $>C=O$ (ketones) | 562 |
| $>CH-$ | 57 | –COO– (esters) | 634 |
| $>C<$ | –190 | –CN | 839 |
| $=CH_2$ | 388 | –H | 164–205 |
| $=CH-$ | 227 | –Cl | 552 |
| $=C<$ | 39 | $-CF_2-$ | 307 |
| Phenyl | 1500 | $-CF_3$ | 561 |
| Phenylene | 1350 | $>Si<$ (silicones) | –77 |

[a] Small PA (1953) J Appl Chem 3: 71; see also: Grulke EA (1989) In: Brandrup J, Immergut EH (eds) Polymer handbook, 3rd edn. Wiley, New York, p VII/519

[b] $D$ is the density of the polymer in $g/cm^3$ and $M_0$ is the molecular weight of the repeat unit in g/mol.

*Examples:*

**1. Poly(styrene)** ($D = 1.05\ g/cm^3$; $M_0 = 104$ g/mol):

| Group | F |
|---|---|
| $-CH_2-$ | 272 |
| $>CH-$ | 57 |
| Phenyl | 1500 |
| $\Sigma F$ | 1829 |

$\delta = (1.05/104) \cdot 1829\ (J/cm^3)^{1/2}$

$\delta = 18.5\ (J/cm^3)^{1/2}$

**2. Poly(carbonate)** [Poly(oxycarbonyloxy-1,4-phenylene-isopropylene-1,4-phenylene)] ($D = 1.20\ g/cm^3$; $M_0 = 254$ g/mol):

| Group | F | Group | F |
|---|---|---|---|
| –O– | 134 | $-CH_3$ | 437 |
| –COO– | 634 | $-CH_3$ | 437 |
| Phenylene | 1350 | Phenylene | 1350 |
| $>C<$ | –190 | $\Sigma F$ | 4152 |

$\delta = (1.20/254) \cdot 4152\ (J/cm^3)^{1/2} = 19.6\ (J/cm^3)^{1/2}$

**Table A 2.** Solubility parameters ($\delta$) and physical constants of solvents

| Solvent | $\delta^{a}$ $(J/cm^3)^{1/2}$ | $D^{25b}$ $g/cm^3$ | $n_{546}^{25b}$ | b.p.[b] °C |
|---|---|---|---|---|
| Acetic acid | 20.7 | 1.0437 | 1.3713 | 118.1 |
| Acetic anhydride | 21.1 | $1.087^{15c}$ | $1.389_D^{25c}$ | 140.0[c] |
| Acetone | 20.3 | 0.7846 | 1.3581 | 56.2 |
| Acetonitrile | 24.3 | 0.7766 | 1.3429 | 81.6 |
| Benzene | 18.8 | 0.8737 | 1.5020 | 80.1 |
| Benzyl alcohol | 24.8 | 1.0416 | 1.5424 | 205.5 |
| Bromobenzene | 20.3 | 1.4879 | 1.5616 | 156.2 |

**Table A 2.** (Continued)

| Solvent | $\delta^a$ $(J/cm^3)^{1/2}$ | $D^{25b}$ $g/cm^3$ | $n_{546}^{25\,b}$ | b.p.$^b$ °C |
|---|---|---|---|---|
| *iso*-Butanol | 21.5 | 0.7982 | 1.3954 | 107.9 |
| *n*-Butanol | 23.3 | 0.8057 | 1.3988 | 117.8 |
| Butanone | 19.0 | 0.7995 | 1.3780 | 79.6 |
| *n*-Butyl acetate | 17.4 | 0.8764 | 1.3933 | 126.1 |
| *n*-Butyl chloride | 16.6 | 0.8811 | 1.4014 | 78.5 |
| *n*-Butyl ether | 16.0 | $0.772^{15c}$ | $1.401_D^{15c}$ | $140.0^c$ |
| *γ*-Butyrolactone | 25.8 | $1.051^d$ | $1.434_D^{25e}$ | $206.0^c$ |
| Chlorobenzene | 19.4 | 1.1013 | 1.5259 | 131.8 |
| *m*-Cresol | 20.9 | 1.0300 | 1.5430 | 202.2 |
| Cyclohexane | 16.8 | 0.7738 | 1.4253 | 80.8 |
| Cyclohexanol | 23.3 | 0.9455 | 1.4666 | 161.1 |
| Cyclohexanone | 20.3 | 0.9421 | 1.4503 | 155.7 |
| Decahydronaphthalene | 18.0 | 0.8659 | 1.4691 | 187.3 |
| *n*-Decane | 13.5 | 0.7264 | 1.4116 | 174.0 |
| *o*-Dichlorobenzene | 20.5 | 1.3003 | 1.5517 | 180.0 |
| 1,2-Dichloroethane | 18.5 | 1.1680 | 1.4164 | 57.3 |
| Dichloromethane | 19.8 | 1.3167 | 1.4236 | 40.2 |
| Diethylene glycol | 24.8 | $1.128^d$ | $1.445_D^{25e}$ | $245.0^c$ |
| Diethyl ether | 15.1 | 0.7078 | 1.3515 | 34.6 |
| Diisopropyl ether | 14.1 | $0.728^{17c}$ | $1.367_D^{25c}$ | $68.0^c$ |
| Diisopropyl ketone | 16.4 | $0.809^{20c}$ | $1.401_D^{20c}$ | $124.5^c$ |
| *N*,*N*-Dimethyl acetamide | 22.1 | $0.943^{20c}$ | $1.437_D^{22c}$ | $165.5^c$ |
| *N*,*N*-Dimethyl formamide | 24.8 | $0.949^{20f}$ | $1.427_D^{25f}$ | $153.0^f$ |
| Dimethyl sulfoxide | 24.6 | 1.0958 | $1.4773_D^{25}$ | 189.0 |
| 1,4-Dioxane | 20.5 | 1.0280 | 1.4218 | 101.4 |
| Diphenyl ether | 20.7 | $1.071^{25c}$ | $1.581_D^{20c}$ | $252.0^c$ |
| Ethanol | 26.0 | 0.7852 | 1.3612 | 78.2 |
| Ethyl acetate | 18.6 | 0.8945 | 1.3715 | 77.2 |
| Ethylbenzene | 18.0 | 0.8626 | 1.4969 | 136.2 |
| Ethylene carbonate | 30.1 | $0.980^{15c}$ | $1.426_D^{25c}$ | $125.8^c$ |
| Ethylene glycol | 29.9 | 1.1099 | 1.4321 | 197.8 |
| Formamide | 39.3 | 1.1247 | 1.4484 | $105.5^g$ |
| Formic acid | 24.8 | 1.2140 | 1.3713 | 100.5 |
| Glycerol | 33.8 | $1.261^{20c}$ | $1.475_D^{20c}$ | $290.0^c$ |
| *n*-Heptane | 15.1 | 0.6795 | 1.3867 | 98.5 |
| *n*-Hexane | 14.9 | 0.6549 | 1.3737 | 68.8 |
| Methanol | 29.7 | 0.7866 | 1.3284 | 64.7 |
| Methyl acetate | 19.6 | 0.9273 | 1.3622 | 57.3 |
| Methyl-*n*-amyl ketone | 17.4 | 0.8111 | 1.4085 | 151.5 |
| Methyl isobutyl ketone | 17.2 | 0.7960 | 1.3959 | 116.1 |
| Nitrobenzene | 20.5 | 1.1983 | 1.5566 | 210.8 |
| Nitroethane | 22.7 | $1.053^{20c}$ | $1.390_D^{25c}$ | $114.0^c$ |
| Nitromethane | 26.0 | 1.1311 | 1.3818 | 101.2 |
| Phenol | 24.8 | $1.0708^c$ | $1.541_D^{40c}$ | $182.2^c$ |
| *iso*-Propanol | 23.5 | 0.7809 | 1.3772 | 82.3 |
| *n*-Propanol | 24.3 | 0.7999 | 1.3847 | 97.2 |
| Pyridine | 21.9 | 0.9778 | 1.5115 | 115.5 |

**Table A 2.** (Continued)

| Solvent | $\delta^a$ $(J/cm^3)^{1/2}$ | $D^{25b}$ $g/cm^3$ | $n_{546}^{25\ b}$ | b.p.[b] °C |
|---|---|---|---|---|
| 1,1,2,2-Tetrachloroethane | 19.8 | 1.5869 | 1.4939 | 146.2 |
| Tetrachloromethane | 17.6 | 1.5844 | 1.4596 | 76.8 |
| Tetrahydrofurane | 18.6 | 0.883 | 1.4066 | 65.0 |
| 1,2,3,4-Tetrahydronaphthalene | 19.4 | 0.9632 | 1.4979 | 207.6 |
| Toluene | 18.2 | 0.8623 | 1.4980 | 110.7 |
| Trichloroethane | 19.6 | 1.4355 | 1.4702 | 113.5 |
| Trichloromethane | 19.0 | 1.4799 | 1.4446 | 61.2 |
| Water | 47.9 | 0.9970 | 1.3340 | 100.0 |
| *p*-Xylene | 18.0 | 0.8567 | 1.4971 | 138.4 |

[a] See: Grulke EA (1989) In: Brandrup J, Immergut EH (eds) Polymer handbook, 3rd ed. John Wiley, New York, p VII/519

[b] See: Johnson BL, Smith J (1972) In: Huglin MB (ed) Light scattering from polymer solutions. Academic, London, p 27; values of density $D^{25}$ and refractive index $n_{546}^{25}$ refer to 25 °C and 546 nm (wavelength of light); b.p. is related to 760 Torr (101.3 kPa). Deviating conditions are specified.

[c] See: Dimroth K (1983) In: D'Ans, Lax – Taschenbuch für Chemiker und Physiker, 4th edn, vol II (Synowietz C, ed). Springer, Berlin, Heidelberg, New York, p **2**-3

[d] See: Elias HG (1989) In: Brandrup J, Immergut EH (eds) Polymer handbook, 3rd edn. John Wiley, New York, p III/13

[e] See: Elias HG (1989) In: Brandrup J, Immergut EH (eds) Polymer handbook, 3rd edn. John Wiley, New York, p III/25

[f] See: Fleischer D (1989) In: Brandrup J, Immergut EH (eds) Polymer handbook, 3rd edn. John Wiley, New York, p III/29

[g] Value at 11 Torr (1.47 kPa)

**Table A 3.** Solvents and nonsolvents for polymers[a] ordered according to solubility parameters $\delta^b$ (cf. Sect. 3.1; differences $\Delta\delta = |\delta_S - \delta_P|$ [cf. Eq. (3.7)] *more* than 2 $(J/cm^3)^{1/2}$ for *solvents* and *less* than 2 $(J/cm^3)^{1/2}$ for *nonsolvents*, respectively, are indicated by *italic* types.)

| Polymer<br>$D^c$<br>$\delta^{b,e}$<br>$\delta$ (Small)[d,e] | Solvents $(\delta_S)^e$ | Nonsolvents $(\delta_S)^e$ |
|---|---|---|
| **Poly(isobutylene)** | *n*-hexane (14.9) | ethyl acetate (18.6) |
| $D = 0.92$ | *n*-heptane (15.1) | butanone (19.0) |
| $\delta$ = **16.0–16.6** | *n*-butyl ether (16.0) | acetone (20.3) |
| $\delta$ (Small) = 15.7 | butyl acetate (17.4) | acetic acid (20.7) |
| | THF (18.6) | ethanol (26.0) |
| | *benzene* (18.8) | nitromethane (26.0) |
| | *trichloromethane* (19.0) | methanol (29.7) |
| | *chlorobenzene* (19.4) | |
| | *dichloromethane* (19.8) | |

**Table A 3.** (Continued)

| Polymer<br>$D^c$<br>$\delta^{b,e}$<br>$\delta$ (Small)$^{d,e}$ | Solvents ($\delta_S$)$^e$ | Nonsolvents ($\delta_S$)$^e$ |
|---|---|---|
| **Poly(ethylene)**<br>$D = 0.92$ (PE-LD)<br>0.96 (PE-HD)<br>$\delta$ = **16.0–16.8**<br>$\delta$ (Small) = 18.0 (LD)<br>18.6 (HD) | decahydronaphthalene (18.0) [70 °C-PE-LD, 135 °C-PE-HD]<br>*p*-xylene (18.0) [75 °C-PE-LD, 100 °C-PE-HD]<br>toluene (18.2) [LD]<br>tetrahydronaphthalene (19.4) [120 °C] | *methyl isobutyl ketone* (17.2)<br>butanone (19.0)<br>acetone (20.3)<br>*n*-propanol (24.3)<br>nitromethane (26.0)<br>methanol (29.7) |
| **Poly(isoprene)-1,4-*cis***<br>$D = 0.91$ (unvulc.)<br>$\delta$ = **16.4–16.8**<br>$\delta$ (Small) = 16.7 | toluene (18.2)<br>THF (18.6)<br>trichloromethane (19.0)<br>dichloromethane (19.8) | *n-hexane* (14.9)<br>*diethyl ether* (15.1)<br>acetone (20.3)<br>acetic acid (20.7)<br>ethanol (26.0)<br>methanol (29.7) |
| **Poly(butadiene)**<br>$D = 0.92$<br>$\delta$ = **16.6–17.6**<br>$\delta$ (Small) = 17.0 | *n*-heptane (15.1)<br>cyclohexane (16.8)<br>butyl acetate (17.4)<br>THF (18.6)<br>benzene (18.8)<br>butanone (19.0)<br>trichloromethane (19.0)<br>*dichloromethane* (19.8) | *n*-propanol (24.3)<br>ethanol (26.0)<br>methanol (29.7)<br>water (47.9) |
| **Poly(chloroprene)**<br>$D = 1.23$ (unvulc.)<br>$\delta$ = **16.6–18.9**<br>$\delta$ (Small) = 19.0 | ethyl acetate (18.6)<br>THF (18.6)<br>benzene (18.8)<br>trichloromethane (19.0)<br>chlorobenzene (19.4)<br>cyclohexanone (20.3)<br>dioxane (20.5)<br>*pyridine* (21.9) | *n-hexane* (14.9)<br>*acetone* (20.3)<br>ethanol (26.0)<br>methanol (29.7)<br>water (47.9) |
| **Poly(styrene)**<br>$D = 1.05$<br>$\delta$ = **17.4–19.3**<br>$\delta$ (Small) = 18.5 | cyclohexane (16.8) [>35 °C]<br>tetrachloromethane (17.6)<br>ethylbenzene (18.0)<br>toluene (18.2)<br>ethyl acetate (18.6)<br>THF (18.6)<br>butanone (19.0)<br>trichloromethane (19.0)<br>dichloromethane (19.8)<br>cyclohexanone (20.3)<br>dioxane (20.5)<br>*DMF* (24.8) | *n*-hexane (14.9)<br>diethyl ether (15.1)<br>*acetone* (20.3)<br>*acetic acid* (20.7)<br>ethanol (26.0)<br>methanol (29.7) |

**Table A 3.** (Continued)

| Polymer<br>$D^{c}$<br>$\delta^{b,e}$<br>$\delta$ (Small)$^{d,e}$ | Solvents ($\delta_S$)$^{e}$ | Nonsolvents ($\delta_S$)$^{e}$ |
|---|---|---|
| **Poly(α-methylstyrene)**<br>$D = 1.07$<br>$\delta$ (Small) = 18.4 | toluene (18.2)<br>THF (18.6)<br>benzene (18.8) | *n*-hexane (14.9)<br>ethanol (26.0)<br>methanol (29.7)<br>water (47.9) |
| **Poly(propylene)**<br>$D = 0.90–0.94$<br>$\delta$ = **18.8 – 19.2**<br>$\delta$ (Small) = 16.8 | *–atactic*:<br>*diethyl ether* (15.1)<br>*isoamyl acetate* (16.0)<br>cyclohexane (16.8)<br>decahydronaphthalene (18.0)<br>toluene (18.2)<br>benzene (18.8)<br>*–isotactic*:<br>decahydronaphthalene (18.0) [135 °C]<br>*p*-xylene (18.0) [85 °C]<br>tetrahydronaphthalene (19.4) [135 °C] | *butanone* (19.0)<br>*acetone* (20.3)<br>ethanol (26.0)<br>methanol (29.7) |
| **Poly(methyl methacrylate)**<br>$D = 1.19$<br>$\delta$ = **18.4–19.5**<br>$\delta$ (Small) = 18.9 | xylene (18.0)<br>toluene (18.2)<br>1,2-dichloroethane (18.5)<br>ethyl acetate (18.6)<br>THF (18.6)<br>benzene (18.8)<br>butanone (19.0)<br>trichloromethane (19.0)<br>chlorobenzene (19.4)<br>dichloromethane (19.8)<br>acetone (20.3)<br>cyclohexanone (20.3)<br>dioxane (20.5)<br>acetic acid (20.7)<br>isobutanol (21.5) [hot]<br>*nitroethane* (22.7)<br>*formic acid* (24.8) | isopropyl ether (14.1)<br>*n*-hexane (14.9)<br>*cyclohexane* (16.8)<br>*tetrachloromethane* (17.6)<br>ethanol (26.0)<br>methanol (29.7)<br>formamide (39.3) |
| **Cellulose triacetate**<br>$D = 1.31$<br>$\delta$ = **18.8** | THF (18.6)<br>ethyl acetate (18.6)<br>trichloromethane (19.0)<br>trichloroethane (19.6)<br>methyl acetate (19.6)<br>dichloromethane (19.8)<br>dioxane (20.5)<br>*ethylene carbonate* (30.1) | *n*-hexane (14.9)<br>diethyl ether (15.1)<br>*toluene* (18.2)<br>*benzene* (18.8)<br>*chlorobenzene* (19.4)<br>ethanol (26.0)<br>methanol (29.7)<br>water (47.9) |

**Table A 3.** (Continued)

| Polymer<br>$D$[c]<br>$\delta$[b,e]<br>$\delta$ (Small)[d,e] | Solvents ($\delta_S$)[e] | Nonsolvents ($\delta_S$)[e] |
|---|---|---|
| **Poly(vinyl acetate)**<br>$D = 1.19$<br>$\delta =$ **19.1–19.6**<br>$\delta$ (Small) = 19.2 | toluene (18.2)<br>THF (18.6)<br>ethyl acetate (18.6)<br>benzene (18.8)<br>trichloromethane (19.0)<br>butanone (19.0)<br>acetone (20.3)<br>dioxane (20.5)<br>acetic acid (20.7)<br>*acetonitrile* (24.3)<br>*DMSO* (24.6)<br>*DMF* (24.8)<br>*nitromethane* (26.0)<br>*methanol* (29.7) | *n*-hexane (14.9)<br>*decahydronaphthalene* (18.0)<br>cyclohexanol (23.3)<br>propanol [anhydrous] (24.3)<br>water (47.9) |
| **Poly(vinyl chloride)**<br>$D = 1.38$<br>$\delta =$ **19.2–20.3**<br>$\delta$ (Small) = 19.6 | *diisopropyl ketone*<br>(16.4) [low MW]<br>toluene (18.2)<br>[low MW]<br>THF (18.6)<br>butanone (19.0)<br>dichloromethane (19.8)<br>cyclohexanone (20.3)<br>nitrobenzene (20.5)<br>dioxane (20.5) [low MW]<br>*o*-dichlorobenzene<br>(20.5) [low MW]<br>*DMSO* (24.6)<br>*DMF* (24.8) | *n*-hexane (14.9)<br>*ethyl acetate* (18.6)<br>*acetone* (20.3)<br>*acetic acid* (20.7)<br>ethanol (26.0)<br>methanol (29.7)<br>ethylene glycol (29.9) |
| **Poly(carbonate)**<br>$D = 1.20$<br>$\delta$ (Small) = 19.6 | THF (18.6)<br>benzene (18.8)<br>trichloromethane (19.0)<br>dichloromethane (19.8)<br>cyclohexanone (20.3)<br>dioxane (20.5)<br>*m*-cresol (20.9)<br>*DMF* (24.8) | *n*-hexane (14.9)<br>*tetrachloromethane* (17.6)<br>*acetone* (20.3)<br>ethanol (26.0)<br>methanol (29.7)<br>water (47.9) |
| **Poly(oxymethylene)**<br>$D = 1.43$<br>$\delta$ (Small) = 19.8<br>(Soluble only<br>at higher<br>temperatures!) | bromobenzene (20.3)<br>diphenyl ether (20.7)<br>*DMF* (24.8)<br>*benzyl alcohol* (24.8)<br>*γ-butyrolactone* (25.8)<br>*ethylene carbonate* (30.1)<br>*formamide* (39.3) | *ethyl acetate* (18.6)<br>ethanol (26.0)<br>methanol (29.7)<br>water (47.9) |

**Table A 3.** (Continued)

| Polymer<br>$D^{c}$<br>$\delta^{b,e}$<br>$\delta$ (Small)$^{d,e}$ | Solvents ($\delta_S$)$^{e}$ | Nonsolvents ($\delta_S$)$^{e}$ |
|---|---|---|
| **Poly(methacrylate)**<br>$\delta$ = **20.7–20.8** | *toluene* (18.2)<br>*THF* (18.6)<br>*ethyl acetate* (18.6)<br>benzene (18.8)<br>butanone (19.0)<br>trichloromethane (19.0)<br>dichloromethane (19.8)<br>acetone (20.3) | diisopropyl ether (14.1)<br>*n*-hexane (14.9)<br>*n*-heptane (15.1)<br>cyclohexane (16.8)<br>tetrachloromethane (17.6)<br>ethanol (26.0)<br>methanol (29.7) |
| **Cellulose trinitrate**<br>$\delta$ = **21.4** | *methyl amyl ketone* (17.4)<br>*ethyl acetate* (18.6)<br>*THF* (18.6)<br>*trichloromethane* (19.0)<br>methyl acetate (19.6)<br>dichloromethane (19.8)<br>acetone (20.3)<br>cyclohexanone (20.3)<br>nitrobenzene (20.5) | diisopropyl ether (14.1)<br>*n*-hexane (14.9)<br>diethyl ether (15.1)<br>benzene (18.8)<br>ethanol (26.0)<br>methanol (29.7)<br>ethylene glycol (29.9)<br>water (47.9) |
| **Poly(ethylene terephthalate)**<br>*D* = 1.33 (amorphous)<br>$\delta$ = **21.5, 21.9**<br>$\delta$ (Small) = 21.9 | nitrobenzene (20.5)<br>*m*-cresol (20.9)<br>*DMSO* (24.6) [hot]<br>*phenol* (24.8)<br>*trichloro acetic acid* | *n*-hexane (14.9)<br>ethyl acetate (18.6)<br>trichloromethane (19.0)<br>butanone (19.0)<br>*acetone* (20.3)<br>ethanol (26.0)<br>methanol (29.7) |
| **Poly(amides) 6 and 66**<br>*D* = 1.1 (amorphous)<br>$\delta$ (*Hoy*)$^{f}$ = 21.9 | *chlorobenzene* (19.4)<br>[>120 °C]<br>*m*-cresol (20.9)<br>*DMSO* (24.6) [>120 °C]<br>*diethylene glycol*<br>(24.8) [>120 °C]<br>*formic acid* (24.8)<br>*phenol* (24.8)<br>*trifluoro ethanol*<br>*trichloro ethanol*<br>*trichloro acetic acid*<br>*ethylene carbonate* (30.1) | diisopropyl ether (14.1)<br>*n*-hexane (14.9)<br>ethyl acetate (18.6)<br>butanone (19.0)<br>trichloromethane (19.0)<br>*acetone* (20.3)<br>ethanol (26.0)<br>methanol (29.7) |
| **Poly(vinyl alcohol)**<br>$\delta$ = **25.8** | DMSO (24.6) [hot]<br>DMF (24.8)<br>*ethylene glycol* (29.9) [hot]<br>*glycerol* (33.8) [hot]<br>*formamide* (39.3)<br>*water* (47.9) | *n*-hexane (14.9)<br>ethyl acetate (18.6)<br>THF (18.6)<br>butanone (19.0)<br>trichloromethane (19.0)<br>dichloromethane (19.8)<br>acetone (20.3) |

**Table A 3.** (Continued)

| Polymer<br>$D^c$<br>$\delta^{b,e}$<br>$\delta$ (Small)[d,e] | Solvents ($\delta_S$)[e] | Nonsolvents ($\delta_S$)[e] |
|---|---|---|
| | | dioxane (20.5)<br>*ethanol* (26.0)<br>methanol (29.7) |
| **Poly(acrylonitrile)**<br>$D$ = 1.17–1.18<br>$\delta$ = **25.3–26.1 (31.5)**<br>$\delta$ (Small) = 26.1 | *acetic anhydride* (21.1)<br>*N,N-dimethyl acetamid* (22.1)<br>DMSO (24.6)<br>DMF (24.8)<br>$\gamma$-butyrolactone (25.8)<br>*ethylene carbonate* (30.1) | *n*-hexane (14.9)<br>THF (18.6)<br>butanone (19.0)<br>trichloromethane (19.0)<br>dichloromethane (19.8)<br>acetone (20.3)<br>dioxane (20.5)<br>*ethanol* (26.0)<br>methanol (29.7) |

[a] Fuchs O (1989) In: Brandrup J, Immergut EH (eds) Polymer handbook, 3rd edn. John Wiley, New York, p VII/379.

[b] See: Grulke EA (1989) In: Brandrup J, Immergut EH (eds) Polymer handbook, 3rd edn. John Wiley, New York, p VII/519.

[c] $D$ represents the density of polymer given in g/cm$^3$ (quoted from: Brandrup J, Immergut EH (eds) Polymer handbook, 3rd edn. John Wiley, New York, Chap V.

[d] $\delta$ (Small) is the solubility parameter of the polymer calculated according to the increment method proposed by Small and shown in Table A 1.

[e] Solubility parameters are given in (J/cm$^3$)$^{1/2}$.

[f] Calculated with increments proposed by Hoy (see footnote b).

**Table A 4.** Simple characteristics of some soluble polymers (pure substances[a])[b]

| Polymer | Density (g · cm$^{-3}$) | Burning behavior[c] | Typical solvents | Features of the solid and compact polymer |
|---|---|---|---|---|
| Poly(ethylene) | 0.92 (PE-LD)<br>0.95 (PE-HD) | white flame with blue center; burning droplets; smell of paraffin | soluble only at higher temperatures | wax-like surface; scratch test with fingernail; not brittle |
| Poly(propylene) | 0.91 – 0.94 | similar to poly(ethylene) | similar to poly(ethylene) | ivory colour; opaque; scratch test with fingernail fails; not brittle |
| Poly(isobutene) | 0.92 | yellow and sooty flame; smell of sealing-wax | benzine | rubbery; opaque |

**Table A 4.** (Continued)

| Polymer | Density ($g \cdot cm^{-3}$) | Burning behavior[c] | Typical solvents | Features of the solid and compact polymer |
|---|---|---|---|---|
| Poly(styrene) | 1.05 | yellow and sooty flame; burning droplets; sweetish smell | toluene, tetrachloromethane | glassy, transparent; brittle; metallic sound |
| Poly(methyl methacrylate) | 1.18 | small, bright, blue, subsequently yellow flame with crackle; fruity smell; white smoke after quenching | acetone | glassy, bright and clear; hollow sound |
| Poly(carbonate) | 1.20 | dark-yellow, sooty, spitting flame; melting without droplets but with inflation; black crust and phenolic smell after quenching | dichloromethane | hard surface; glassy, bright and clear; full sound; high-impact |
| Poly(ethylene terephthalate) | 1.40 | hard-burning, small, yellow flame; a foil of the material retreats | *m*-cresol | fibers or foils; glassy; strong tightness |
| Poly(oxymethylene) | 1.41 | small, blue, difficult-to-quench flame; droplets; smell of formaldehyde | soluble only at higher temperatures | smooth and hard surface; white and opaque; stiff; rattling sound; not brittle |
| Poly(acrylonitrile) | 1.18 | bright flame; droplets; smell of ammonia and cyanide | DMF | glassy, bright and clear; high-impact |
| Poly(amides) | 1.10 – 1.25 | poorly flammable, small, blue-yellow flame; decomposition; unpleasant smell | *m*-cresol formic acid | ivory colour, milky; tight and tough, not brittle; hollow sound |
| Poly(vinyl chloride) | 1.38 | hard-burning, yellow, sooty, self-quenching flame; no droplets; acrid smell (HCl); Beilstein's test is positive | cyclohexanone, acetone | clear; brittle (cold); rattling sound; stress whitening |

**Table A 4.** (Continued)

| Polymer | Density ($g \cdot cm^{-3}$) | Burning behavior[c] | Typical solvents | Features of the solid and compact polymer |
|---|---|---|---|---|
| Cellulose acetate | 1.31 | yellow, spitting flame; droplets; smell of acetic acid and burned paper | ethyl acetate | brightness; tough; hollow sound |

[a] Additives as fillers, pigments, plasticizers, reinforcing materials etc. can alter these characteristics.
[b] See also: Braun D (1986) Erkennen von Kunststoffen, 2. Aufl (mit der Kunststoff-Bestimmungstafel von H. Saechtling), or: (1982) Simple Methods for Identification of Plastics (with the Plastics Identification Table by H. Saechtling). Carl Hanser, München; Carlowitz B (ed) (1990) Kunststoff Handbuch 1, Die Kunststoffe: Chemie, Physik, Technologie. Carl Hanser, München
[c] Burning behavior must be tested carefully using the following equipment: exhauster, fireproof plate, tweezers; no more than 1 g polymer, rod-like or bar-shaped specimens should be used.

**Table A5.** Generalized Schulz distribution

$$\mathrm{W}(P) = [c\Gamma(k+1)]^{-1}(P/c)^k \cdot \exp(-P/c)$$

$$\mathrm{N}(P) = [c^2\Gamma(k+1)]^{-1}(P/c)^{k-1} \cdot \exp(-P/c)$$

with: $k = 1/U$ (cf. Eq. (2.6))

$$c = \bar{P}_\mathrm{w} - \bar{P}_\mathrm{n} = (\bar{M}_\mathrm{w} - \bar{M}_\mathrm{n})/M_0$$

$\Gamma$ is the tabulated gamma function which for integral values of $k$, yields:

$$\Gamma(k+1) = k\Gamma(k) = k!$$

$$\begin{aligned}\mathrm{W}(M) &= \mathrm{W}(P)/M_0 \\ &= [M_0 c\Gamma(k+1)]^{-1}(M/M_0 c)^k \cdot \exp(-M/M_0 c)\end{aligned}$$

$$\begin{aligned}\mathrm{N}(M) &= \mathrm{N}(P)/M_0 = \mathrm{W}(M)/P \\ &= [M_0 c^2\Gamma(k+1)]^{-1}(M/M_0 c)^{k-1} \cdot \exp(-M/M_0 c)\end{aligned}$$

$$\bar{P}_\mathrm{n} = k \cdot c \qquad \bar{M}_\mathrm{n} = M_0 \cdot k \cdot c$$

$$\bar{P}_\mathrm{w} = (k+1)c \qquad \bar{M}_\mathrm{w} = M_0(k+1)c$$

$\mathrm{I}(P)$ and $\mathrm{I}(M)$ can only be calculated numerically.

**Table A6.** Tung distribution

$$\mathrm{I}(P) = 1 - \exp(-a \cdot P^b) \tag{1}$$

$$\mathrm{W}(P) = a \cdot b \cdot P^{b-1} \cdot \exp(-a \cdot P^b)$$

$$\mathrm{N}(P) = a \cdot b \cdot P^{b-2} \cdot \exp(-a \cdot P^b)$$

$$\mathrm{I}(M) = 1 - \exp(-a \cdot M^b/M_0^b) \tag{2}$$

$$\mathrm{W}(M) = \mathrm{W}(P)/M_0$$

$$= (a/M_0^b) \cdot b \cdot M^{b-1} \cdot \exp(-a \cdot M^b/M_0^b)$$

$$\mathrm{N}(M) = \mathrm{N}(P)/M_0$$

$$= (a/M_0^{b-1}) \cdot b \cdot M^{b-2} \cdot \exp(-a \cdot M^b/M_0^b)$$

$$\bar{P}_n = \{a^{1/b} \cdot \Gamma[1 - (1/b)]\}^{-1} \qquad \bar{M}_n = M_0\{a^{1/b} \cdot \Gamma[1 - (1/b)]\}^{-1}$$

$$\bar{P}_w = \Gamma[1 + (1/b)] \cdot (a^{1/b})^{-1} \qquad \bar{M}_w = M_0 \Gamma[1 + (1/b)] \cdot (a^{1/b})^{-1}$$

$$\bar{P}_w/\bar{P}_n = \bar{M}_w/\bar{M}_n = \Gamma[1 + (1/b)]\Gamma[1 - (1/b)] = U + 1$$

Parameters $a$ and $b$ can be obtained from the logarithmic form of Eqs. (1) and (2):

(1) $\log\{\log[1 - \mathrm{I}(P)]^{-1}\} = b \cdot \log P + \log(a/2.303)$

(2) $\log\{\log[1 - \mathrm{I}(M)]^{-1}\} = b \cdot \log M + \log(a/2.303\, M_0^b)$

**Table A 7.** Logarithmic normal distribution (Wesslau distribution)

$$\mathrm{W}(P) = (\sigma\sqrt{2\pi})^{-1} \cdot P^{-1} \cdot \exp[-\ln^2(P/P^*)/2\sigma^2]$$

$$\mathrm{N}(P) = (\sigma\sqrt{2\pi})^{-1} \cdot P^{-2} \cdot \exp[-\ln^2(P/P^*)/2\sigma^2]$$

with: $P^*$ – abscissa value $P$ corresponding to $\mathrm{I}(P^*) = 0.5$ (median value)

$$P^* = (\bar{P}_w \cdot \bar{P}_n)^{1/2}$$

$\sigma$ – standard deviation of $\ln P$

$$\pm\sigma = \ln(P/P^*) \text{ for } \mathrm{I}(P) = \mathrm{I}(P^*) \pm 0.3413$$

$$\sigma = [\ln(\bar{P}_w/\bar{P}_n)]^{1/2}$$

$$\mathrm{W}(M) = \mathrm{W}(P)/M_0$$

$$= (\sigma\sqrt{2\pi})^{-1} \cdot M^{-1} \cdot \exp[-\ln^2(M/M^*)/2\sigma^2]$$

$$\mathrm{N}(M) = \mathrm{N}(P)/M_0$$

$$= (\sigma\sqrt{2\pi})^{-1} \cdot M_0 M^{-2} \cdot \exp[-\ln^2(M/M^*)/2\sigma^2]$$

with $M^* = M_0 \cdot P^*$

$$\bar{P}_n = P^* \cdot \exp(-\sigma^2/2) \qquad \bar{M}_n = M^* \cdot \exp(-\sigma^2/2)$$

$$\bar{P}_w = P^* \cdot \exp(+\sigma^2/2) \qquad \bar{M}_w = M^* \cdot \exp(+\sigma^2/2)$$

$$\bar{P}_w/\bar{P}_n = \bar{M}_w/\bar{M}_n = \exp\sigma^2 = U + 1$$

$\mathrm{I}(P)$ and $\mathrm{I}(M)$ can only be calculated numerically, but, graphic evaluation by cumulative plotting in a sum probability grid with logarithmic abscissa enables us to obtain median value $P^*$ and standard deviation $\sigma$.

**Table A 8.** Phase relations for simple one-step and multi-step extraction (without chromatographic or counter-current steps)

Number of fractions: 1, 2, . . .
Number of extraction steps per fraction: a, b, . . .
$w_{i,o}$ – mass of polymer with $P_i$ in starting polymer; other symbols see Eqs. (3.8) through (3.11).

A One-step extraction:

$$w_{i,1} = w_{i,o}$$

$$w_{i,1''} = w_{i,o}/[1+(V'/V'')_1 \exp(-k_1 P_i)] \quad \text{(acc. to (3.10))}$$

$$w_{i,1'} = w_{i,o}/[1+(V''/V')_1 \exp(k_1 P_i)] \quad \text{(acc. to (3.11))}$$

$$w_{i,2} = w_{i,1''}$$

$$w_{i,2''} = w_{i,o}/\{[1+(V'/V'')_1 \exp(-k_1 P_i)]* \\ *[1+(V'/V'')_2 \exp(-k_2 P_i)]\}$$

$$w_{i,2'} = w_{i,o}/\{[1+(V'/V'')_1 \exp(-k_1 P_i)]* \\ *[1+(V''/V')_2 \exp(k_2 P_i)]\}$$

$$w_{i,3} = w_{i,2''}$$

. . .

B Multi-step (e.g., Soxhlet) extraction, restricted to two steps:

$$w_{i,1a''} = w_{i,1}'' \quad \text{(see A)}$$

$$w_{i,1a'} = w_{i,1'} \quad \text{(see A)}$$

$$w_{i,1b} = w_{i,1a''}$$

$$w_{i,1b''} = w_{i,o}/[1+(V'/V'')_1 \exp(-k_1 P_i)]^2$$

$$w_{i,1b'} = w_{i,o}/\{[1+(V'/V'')_1 \exp(-k_1 P_i)]* \\ *[1+(V''/V')_1 \exp(k_1 P_i)]\}$$

$$w_{i,2a} = w_{i,1b''}$$

$$w_{i,2a''} = w_{i,o}/\{[1+(V'/V'')_1 \exp(-k_1 P_i)]^2* \\ *[1+(V'/V'')_2 \exp(-k_2 P_i)]\}$$

$$w_{i,2a'} = w_{i,o}/\{[1+(V'/V'')_1 \exp(-k_1 P_i)]^2* \\ *[1+(V''/V')_2 \exp(k_2 P_i)]\}$$

$$w_{i,2b} = w_{i,2a''}$$

. . .

**Table A 9.** Generation of logarithmic solvent/nonsolvent gradients in a stepwise procedure by use of small mixing vessels

The customarily used continuous gradient can be replaced by a stepwise generation using small volumes of differently composed storage mixtures. The calculation follows Eq. (7.1) in Sect. 7.1. Two variants are possible which will be illustrated in two examples.

- *Replacement of a mixing vessel with $V_m = 4$ l by another one with $V_m = 2$ l and use of storage volumes of 2 l*
- Assumed gradient limits (related to the solvent component): $\varphi_{m,0} = 0.140$, $\varphi_{st} = 0.649$
- Calculation of the continuous gradient ($\varphi_{m,t}$) with $V_m = 4$ l in dependence of the eluent volume ($\dot{V} \cdot t$) – see Table A 9.1
- Calculation of $\varphi_{st}$ for each step ($\Delta(\dot{V} \cdot t) = 2$ l) using $V_m = 2$ l, the desired $\varphi_{m,t}$ of the continuous gradient at 2, 4, . . . l, and $\varphi_{m,0}$ of the respective step ($\varphi_{m,t}$ of the previous step) – see Table A9.1
- With storage volumes of 2 l in each step having the composition $\varphi_{st}$, values $\varphi_{m,t}$ ($V_m = 2$ l) can be calculated. Of course, intermediate values of $\dot{V} \cdot t$ can be obtained – see Table A 9.1.
- The difference in $\varphi_{m,t}$ between continuous and stepwise gradient is given in Table A 9.1. The gradients are plotted in Fig. A 9a. A reasonable approximation can be stated.

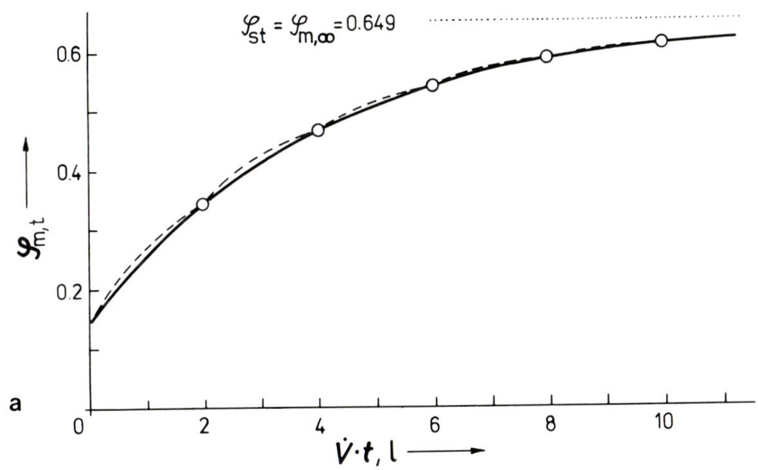


**Fig. A 9a.** Continuous (———) and stepwise adapted (– – –) solvent/nonsolvent gradients. Calculation according to Table A 9.1 using consecutive storage volumes of 2 l each; $V_m = 2$ l (– – –) instead of 4 l (———); ○ – change of storage volumes for stepwise adaption, · · · · – storage composition by use of the continuous gradient

- *Replacement of a mixing vessel with $V_m = 2.7$ l by another one with $V_m = 0.3$ l and use of different storage volumes*

A simplified version of stepwise gradient generation is described which requires, however, smaller storage volumes than used in the first example.

- Assumed gradient limits (related to the solvent component): $\varphi_{m,0} = 0.5$, $\varphi_{st} = 0.8$
- Calculation of the continuous gradient ($\varphi_{m,t}$) with $V_m = 2.7$ l in dependence of the eluent volume ($\dot{V} \cdot t$) – see Table A 9.2

**Table A 9.** (Continued)

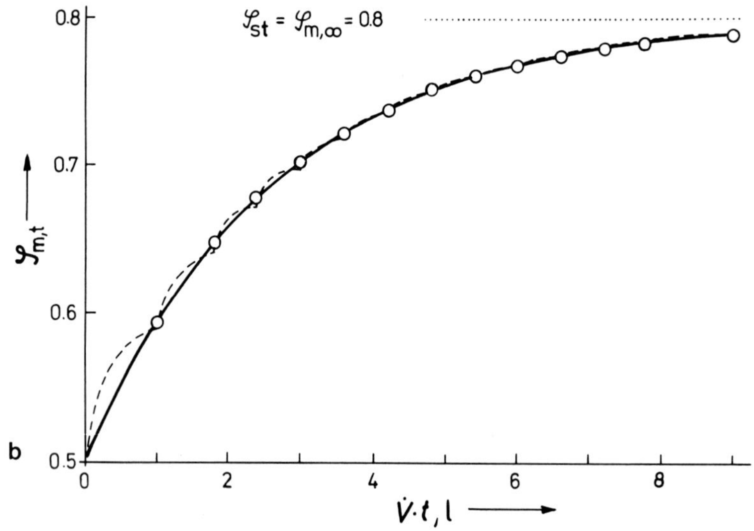


**Fig. A 9b.** Continuous (———) and stepwise adapted (– – –) solvent/nonsolvent gradients. Calculation according to Table A 9.2 using varying consecutive storage volumes; $V_m = 0.3$ l (– – –) instead of 2.7 l (———); ○ – change of storage volumes for stepwise adaption, · · · · – storage composition by use of the continuous gradient

- The gradient is approximated by consecutive filling of the storage vessel with different volumes (1 l, 0.8 l, 0.6 l, etc.) of those compositions which follow from the continuous gradient (values in Table A 9.2).
- The stepwise gradient can be calculated by use of $V_m = 0.3$ l, $\varphi_{st} = \varphi_{m,t}$ (Table A 9.2) in each step, $\varphi_{m,0} = \varphi_{m,t}$ calculated for the previous step (not given here), and $\dot{V} \cdot t = 1$ l, 0.8 l, 0.6 l, etc. The result is little different from the first example since here the deviations are both positive and negative.
- The result is depicted in Fig. A 9b. It turns out to be a reasonable approximation. Of course, the gradient can be subdivided into other volume steps.

**Table A 9.1.** Stepwise generation of solvent/nonsolvent gradients, calculated according to Eq. (7.1)

| $\dot{V}\cdot t$ (l) | $\varphi_{m,t}$ [a] $V_m = 4$ l [b] | $V_m = 2$ l [c] | Deviation (%) | Step [c] | $\varphi_{m,0}$ [d] | $\varphi_{st}$ [c,e] |
|---|---|---|---|---|---|---|
| 0 | 0.140 | 0.140 | 0 | | 0.140 | |
| 1 | 0.253 | 0.266 | 5.3 | I | | |
| 2 | 0.340 | 0.342 | 0.6 | | | 0.460 |
| | | | | | 0.342 | |
| 3 | 0.409 | 0.417 | 2.0 | II | | |
| 4 | 0.462 | 0.462 | 0 | | | 0.532 |
| | | | | | 0.462 | |
| 5 | 0.503 | 0.509 | 1.2 | III | | |
| 6 | 0.536 | 0.538 | 0.4 | | | 0.582 |
| | | | | | 0.538 | |
| 7 | 0.561 | 0.566 | 0.9 | IV | | |
| 8 | 0.580 | 0.582 | 0.3 | | | 0.608 |
| | | | | | 0.582 | |
| 9 | 0.596 | 0.598 | 0.4 | V | | |
| 10 | 0.607 | 0.607 | 0 | | | 0.622 |
| $\infty$ | 0.649 | | | | | |

[a] $\varphi_{m,t}$ is here related to the volume of the eluent ($\dot{V}\cdot t$) and denotes the solvent content.
[b] Continuous gradient using $\varphi_{st} = 0.649$ ($\dot{V}\cdot t = \infty$)
[c] Stepwise logarithmic gradient produced in steps I to V with respective 2 l storage mixture having $\varphi_{st}{}^{e}$
[d] Starting compositions $\varphi_{m,0}$ of steps I to V are the final compositions $\varphi_{m,t}$ of the foregoing steps.
[e] Values $\varphi_{st}$ are calculated using the values $\varphi_{m,0}{}^{d}$ of each step and the desired value $\varphi_{m,t}$ according to Eq. (7.1). Example for step I: $\varphi_{m,0}{}^{I} = 0.140$, $\varphi_{m,t}{}^{I} = 0.340$, $\dot{V}\cdot t = 2$ l, $V_m = 2$ l. Rearrangement of Eq. (7.1) yields $\varphi_{st}{}^{I} = 0.457$. $\varphi_{st}{}^{I} = 0.460$ was used leading to $\varphi_{m,2} = 0.342$.

**Table A 9.2.** Continuous solvent/nonsolvent gradient calculated according to Eq. (7.1) with $V_m = 2.7$ l

| $\dot{V}\cdot t$ (l) | $\varphi_{m,t}$ | $\dot{V}\cdot t$ (l) | $\varphi_{m,t}$ | $\dot{V}\cdot t$ (l) | $\varphi_{m,t}$ | $\dot{V}\cdot t$ (l) | $\varphi_{m,t}$ |
|---|---|---|---|---|---|---|---|
| 0 | 0.500 | 3.0 | 0.701 | 5.4 | 0.759 | 7.8 | 0.783 |
| 1.0 | 0.593 | 3.6 | 0.721 | 6.0 | 0.767 | 9.0 | 0.789 |
| 1.8 | 0.646 | 4.2 | 0.737 | 6.6 | 0.774 | $\infty$ | 0.800 |
| 2.4 | 0.677 | 4.8 | 0.749 | 7.2 | 0.779 | | |

**Table A 10.** Compilation of fractionation procedures in the literature

*Different techniques and polymers:*

Cantow MJR (ed) (1967) Polymer fractionation. Academic, New York (survey especially in Chapter G)
Tung LH (ed) (1977) Fractionation of synthetic polymers. Marcel Dekker, New York
Bello A, Barrales-Rienda JM, Guzman GM (1989) In: Brandrup J, Immergut EH (eds) Polymer handbook, 3rd edn. John Wiley, New York, p VII/233
Theil J, Calugaru EM, Feldman D (1972) Faserforsch Textiltechn 23: 123 (polyamides and polyesters)

*Copolymers:*

Fuchs O, Schmieder W (1967) In: Cantow MJR (ed) Polymer fractionation. Academic, New York, p 341
Riess G, Callot P (1977) In: Tung LH (ed) Fractionation of synthetic polymers. Marcel Dekker, New York, p 445

*Column fractionation:*

Barrel EM, Johnson JF, Cooper AR (1977) In: Tung LH (ed) Fractionation of synthetic polymers. Marcel Dekker, New York, p 267

*Baker-Williams fractionation:*

Porter SR, Johnson JF (1967) In: Cantow MJR (ed) Polymer fractionation. Academic, New York, London, p 95
Glöckner G (1987) Polymer characterization by liquid chromatography. Elsevier, Amsterdam, p 467

*TREF:*

Wild L (1990) Adv Polym Sci 98: 1
Glöckner G (1990) J Appl Polym Sci, Appl Polym Symp 45: 1
Wild L (1993) Trends in Polym Sci 1: 50

*Fractionation with demixing solvents:*

Kuhn R (1976) Makromol Chem 177: 1525
Kuhn R (1983) In: Klempner D, Frisch KC (eds) Polymer alloys III. Plenum, New York, p 45 (cited as Table 11.6 in this book)

# Glossary of Terms

Binodal
Branching of polymers
Cloud point
Coexistence curve
Copolymers
Azeotropic copolymer, block copolymer, chemically heterogeneous copolymer, chemically homogeneous copolymer, graft copolymer, quasi-copolymer, statistical copolymer
Craig partition
Critical point
Gamma function
Glass-transition temperature
Interstitial volume
Intrinsic viscosity
Kuhn-Mark-Houwink-Sakurada equation
LCST
Theta state
UCST

## Binodal

The binodal summarizes the points of composition of coexistent phases, generated by decomposition of liquid mixtures (solutions), independently of the composition of the initially homogeneous system. Conjugated phases, having $\mu'_P = \mu''_P$, are joined by a tie line.

In the case of genuine binary or ternary mixtures (i.e., none of the components is a mixture of several species), binodal, coexistence curve, and turbidity (cloud-point) curve coincide.

In quasi-binary or quasi-ternary systems (i.e., one of the constituents is a mixture of polymerhomologous species having different DPs), different coexistence curves of the phases, which are no binodals, exist in dependence on composition. Cloud-point curve and coexistence curves differ from another.

## Branching of Polymers

Branched polymer structures are caused by chain-transfer reactions, radiation, and special syntheses. The solution behavior of branched molecules is more or less different from linear molecules of the same chemical composition and molecular weight. Branched molecules yield smaller values of intrinsic viscosity and SEC elution volume than corresponding linear molecules.

## Cloud Point

The cloud point of a mixed system marks the appearance of the first turbidity in an initially clear solution due to the onset of phase separation. In a polymer solution, the new phase is the gel phase. Cloud points can be determined by cooling the solution (yielding the cloud-point temperature) or addition of nonsolvent to the solution (resulting in the volume fraction of nonsolvent, $\varphi_{NS}$). A graph, representing cloud-point temperatures or $\varphi_{NS}$ values as a function of composition of the solution, is called the cloud-point (turbidity) curve which is largely symmetric for low-molecular mixtures (e.g., two solvents) but extremely skewed for polymer solutions. Cloud-point curves of polymer solutions are strongly influenced by molecular weight and molecular weight distribution of the polymer.

Only for genuine binary and ternary mixtures, cloud-point curve, coexistence curve of the phases, and binodal are identical.

Notice that cloud point and precipitation point in turbidimetric titration ($\varphi^*$) coincide only in the case of sharp fractions (cf. Fig. 5.4, curve (3))!

## Coexistence Curve

Coexistent phases in the thermodynamic equilibrium lie on the coexistence curve. This curve is a binodal only in that case when all points of conjugated phases result in a single curve independently of the starting mixture. This is possible only in genuine binary and ternary mixtures.

In real polymer solutions (quasi-binary or quasi-ternary), coexistence curves depend on polymer concentration due to the influence of molecular weight distribution (i.e., fractionation efficiency represented by the compositions of sol and gel phase depends on polymer concentration). The latter also causes the difference in cloud-point and coexistence curve. Cloud points are related usually to unfractionated polymers whereas coexisting sol and gel phases are the result of a fractionation step due to phase separation.

## Copolymers

Copolymers are synthesized from at least two different monomeric units and can be characterized by overall composition (expressed as mass fraction $w_i$ or mole fraction $x_i$ of the monomeric units i), chemical-composition heterogeneity, and sequence of monomeric units in the copolymer molecules.

An *azeotropic copolymer* is a special type of (mostly binary) statistical copolymers synthesized from the so-called azeotropic monomer mixture having the monomer ratio (mole fraction $x_i$, i = A, B) $x_A/x_B = (r_B - 1)/(r_A - 1)$. $r_A$ and $r_B$ are the monomer reactivity ratios (homopolymerization/copolymerization) related to the monomeric units A and B, respectively. Azeotropic monomer compositions are possible only for $r_A < 1$ and $r_B < 1$ (or, practically irrelevant, $r_A > 1$ and $r_B > 1$). An azeotropic copolymer has, independently of conversion, the composition of the monomer mixture and no conversion heterogeneity of composition.

A *block copolymer* is composed of different blocks with extended sequences of respective only one monomeric species. The number of blocks is usually denoted as diblock, triblock, . . . copolymer, e.g., a block of A units is followed by a block of B units in a diblock copolymer. Block copolymers are synthesized by special techniques, e.g., anionic polymerization.

A *chemically heterogeneous copolymer* is synthesized, apart from azeotropic mixture, to high degrees of conversion without manipulation of the monomer mixture. Resulting conversion heterogeneity is due to the change of $x_A/x_B$ in the monomer mixture with increasing conversion leading to different compositions of the molecules at different conversions.

A *chemically homogeneous copolymer* is synthesized from monomer mixture whose composition remains constant during progressive conversion (what is possible with azeotropic mixtures, in the case of $r_A = r_B = 1$, or by manipulation of the mixture). Such a copolymer possesses only statistical heterogeneity, caused by the statistics of monomer addition in the propagation reaction. This heterogeneity becomes negligible for high degrees of polymerization. Conversion heterogeneity does not occur.

In a *graft copolymer*, backbone and side chains have different chemical constitution. Graft copolymers are synthesized by special branching reactions.

A *quasi-copolymer* originates by a polymeranalogous reaction. The sequence of chemically different units depends on reaction conditions and can be random (polymerization statistics does not work!) or block-shaped.

A *statistical copolymer* has a monomer sequence according to the reaction statistics directed by $r_A$ and $r_B$ values. When both $r_A$ and $r_B$ values tend toward zero, the statistical copolymer changes into an *alternating* copolymer with $x_A/x_B = 1$ and alternate monomeric units.

## Craig Partition

Craig partition is a liquid–liquid partition using a large number of separating funnels which are connected in series. A substance mixture, for instance a polymer with polymerhomologous constituents of different DP, is distributed according to the partition coefficients of the species. Substance exchange and phase transport are carried out in successive, repeating cycles. Substance exchange is reached simultaneously in all vessels by shaking followed by simultaneous transfer of usually the upper phases into the next subsequent vessels. Then, the next exchange step takes place, etc. Usually, the extracting agent in all separating funnels has the same composition, i.e., partition works with constant coefficients of the species.

## Critical Point

At the critical point, coexisting phases coincide and merge in the critical phase. This is possible at the critical temperature and the critical composition (concentration) of the mixture. The critical point is situated on the binodal or the coexistence curve; the tie line changes in this point from a secant to a tangent. In genuine binary systems, the critical point is the vertex of the binodal and represents the maximum cloud-point temperature.

In quasi-binary systems (e.g., polymer with molecular weight distribution dissolved in a single solvent), the critical point results from point of intersection of the turbidity curve and the coexistence curve having the critical concentration (only a mixture of this concentration intersects the cloud-point curve). This point lies below the maximum of the turbidity curve.

In a ternary solvent system at constant pressure and temperature, the critical point usually differs from the vertex of the binodal dependent on the directions of tie lines. The latter are ruled by the values of the Huggins constant $\chi_{ij}$ related to the three components ($\chi_{12}$, $\chi_{13}$, $\chi_{23}$).

In a ternary system at constant pressure and temperature consisting of two solvents and a polymer having a molecular weight distribution (quasi-ternary system), the critical point follows from point of intersection of turbidity and coexistence curve.

Critical points can be extrapolated via centres of tie lines.

## Gamma Function

$$\Gamma(a+1) = \int_0^\infty \exp(-t) \cdot t^a dt$$

which leads for positive integer a to $\Gamma(a+1) = a\Gamma(a) = a!$

## Glass-Transition Temperature ($T_g$)

Amorphous polymers possess a characteristic glass-transition temperature which separates the glassy state (below $T_g$) from plastic behaviour (above $T_g$). Motion of macromolecules against each other is possible only above $T_g$ whereas below this temperature, only segmental motions inside the macromolecules take place. Glass-transition temperature leads to a kink in a plot of bulk volume vs. temperature. Diffusion processes inside the polymer material are hindered in the glassy state.

## Interstitial Volume

The total volume of a packed column is divided into the volume which is occupied by the particles of the packing material and in the interstitial volume amongst the particles. The latter is nearly 26% of the total volume in the case of the closest spherical packing. For real columns with packings less dense than the closest packing, the interstitial volume is larger than 26% of the total volume.

## Intrinsic Viscosity ([$\eta$])

Extrapolation of the ratio $\eta_{sp}/c$, measured with polymer solutions of graded concentration, at concentration $c = 0$ and zero shear rate yields the limiting value $[\eta]$ with the dimension volume/mass (mostly ml/g). $[\eta]$ can be used as measure of the molecular weight via the Kuhn-Mark-Houwink-Sakurada equation.

## Kuhn-Mark-Houwink-Sakurada (KMHS) Equation

The general form of this equation reads

$$[\eta] = K_v \cdot \bar{M}_v^a$$

with $\bar{M}_v$ as viscosity average of the molecular weight. $K_v$ and $a$ are empirical constants which can be determined by calibration with polymerhomologous fractions of known molecular weight and structure. $a$ lies, for linear molecules, in the range from 0.5 (theta state) to 1; branching diminishes $a$. KMHS equation enables the calculation of $\bar{M}_v$ from experimental $[\eta]$. Note the influence of branching on $[\eta]$ and $a$!

## LCST (Lower Critical Solution Temperature)

Systems with LCST behavior display phase separation at elevated temperatures above the critical solution temperature. The latter, plotted vs. mixing ratios of the components, leads to a concave demixing curve. Generally, phase decomposition results in a polymer-rich and a polymer-poor phase. The differences in polymer concentrations of the phases are reduced with descending temperature and finally, disappear at the critical temperature.

LCST behavior is ruled by the thermodynamic condition (see Eq. (3.1) and cf. UCST) $|T\Delta S_{mix}| > |\Delta H_{mix}|$ while solubility below the critical temperature occurs when $|T\Delta S_{mix}| < |\Delta H_{mix}|$ where both quantities, $\Delta H_{mix}$ and $\Delta S_{mix}$, are negative. $\Delta H_{mix} < 0$ is caused by strongly polar interactions between the components (Eq. (3.2) does not hold in this case); $\Delta S_{mix} < 0$ is connected with the formation of ordered structures within the solution. LCST behavior is found often in aqueous polymer solutions.

## Theta State

Polymer solutions show quasi- or pseudo-ideal behavior in the theta state. This is, in a certain solvent or in a solvent mixture (theta solvent), only possible at a single, thermodynamically defined temperature $T = \Delta H_{mix}/\Delta S_{mix}$ ($\Delta G_{mix} = 0$, see Eq. (3.1)), called theta temperature. The theta state marks the limit of phase stability of a polymer solution (with $P \rightarrow \infty$). Some parameters have characteristic values in the theta state, e.g., $\chi = 0.5$ ($P \rightarrow \infty$) and $a = 0.5$ (KMHS equation).

## UCST (Upper Critical Solution Temperature)

Thermally induced phase separation upon cooling below the critical solution temperature is called UCST behavior. The critical temperature, plotted vs. mixing ratios of the components, shows a convex demixing curve. Phase separation leads, with respect to polymer concentration, to the coexistence of a highly diluted and a concentrated phase. The compositions of the phases approach, in contrast to LCST, with ascending temperature and finally, merge in the homogeneous state at the critical temperature.

UCST behavior is governed by the thermodynamic condition (see Eq. (3.1) and cf. LCST) $T\Delta S_{mix} < \Delta H_{mix}$ whereas solubility occurs for $T\Delta S_{mix} > \Delta H_{mix}$ with both quantities, $\Delta H_{mix}$ and $\Delta S_{mix}$, being positive.

# Subject Index

Italic numbers indicate figures and tables.

# Springer-Verlag and the Environment

We at Springer-Verlag firmly believe that an international science publisher has a special obligation to the environment, and our corporate policies consistently reflect this conviction.

We also expect our business partners – paper mills, printers, packaging manufacturers, etc. – to commit themselves to using environmentally friendly materials and production processes.

The paper in this book is made from low- or no-chlorine pulp and is acid free, in conformance with international standards for paper permanency.